Axiomatic Theory of Economics

AXIOMATIC THEORY OF ECONOMICS

VICTOR AGUILAR

Nova Science Publishers, Inc.
Commack, New York

Editorial Production: Susan Boriotti
Office Manager: Annette Hellinger
Graphics: Frank Grucci and John T'Lustachowski
Information Editor: Tatiana Shohov
Book Production: Donna Dennis, Patrick Davin, Christine Mathosian and Tammy Sauter
Circulation: Maryanne Schmidt
Marketing/Sales: Cathy DeGregory

Library of Congress Cataloging-in-Publication Data
available upon request

ISBN 1-56072-296-7

6080 Jericho Turnpike, Suite 207
Commack, New York 11725
Tele. 516-499-3103 Fax 516-499-3146
e-mail: Novascience@earthlink.net
e-mail: Novascil@aol.com
Web Site: http://www.nexusworld.com/nova

Printed in the United States of America

CONTENTS

PREFACE

This book is about a new economic theory. It is not a simplified version of mainstream economics. It does not predict the future, calling neither prosperity or ruin in America. It is certainly not in the "how to be a salesman" genre, nor does it propose to tell the reader how to make money in the framework of current financial institutions. It is an abstract treatise. The purpose of this book is to give an axiomatic† foundation for the theory of economics. The success of the axiomatic method employed by Euclid (in geometry), Kolmogorov (in probability), and others is well known and I claim that similar success can be realized in economics. However, by defining economics to be concerned with the creation of wealth rather than the allocation of scarce resources, I have not only solidified it but have shifted its basic paradigm. I address the issue of price and stock. Supply and demand does not work. This is a fundamental departure from mainstream economics comparable to that of Copernicus in astronomy.

I assert that only a complete reworking of economic theory will assure the advancement of western civilization. The new theory of economics must start from the ground up, with the rejection of supply

† ***Axiom***: a proposition that is assumed without proof for the sake of studying the consequences that follow from it. One's axioms are introduced at the beginning of one's book and then all of the subsequent theorems are derived from the axioms. For example, Euclid claimed that, given a line and a point not on the line, there exists a unique line that passes through the point and is parallel to the given line. Lobachevski asserted that there is more than one parallel and Riemann that there are none. Thus, there are three geometries and, similarly, there may be more than one economics. I am, however, the only economist to ever precisely state my axioms and to base an entire theory on exactly those axioms and on nothing else.

and demand in favor of price and stock, and it must begin with well-considered axioms from which theorems can be deduced in true mathematical sense. It is not sufficient to pay lip-service to the axiomatic method and insert complicated looking equations into journal articles; equations that have just been gleaned from a mathematics or physics textbook. If western civilization is to survive, there must be a revolution in economic theory, and it must be a far more fundamental one than the so-called revolution of Keynes. Economists themselves are incapable of initiating change. It is impossible in the context of writing ten-page long federally-funded journal articles, which is their only job. If such a revolution is to occur, it must come from within pure mathematics. I predict that this revolution will occur, and for this reason alone do I argue against the decline of western civilization.

The potential in the hard sciences is staggering. Not enough credit can be given to engineers for developing modern technology and for doing it in spite of the MBAs that they usually have to work under. If it were not for engineers, economists would have ruined us long ago. But without a new theory of economics, recessions will turn into depressions and the western world will slide back into the dark ages even as engineers and biologists continue to delve into the innermost secrets of life and the obstacles that it faces on this planet.

The mathematical maturity assumed exceeds that of most professional economists (three semesters of calculus, a semester of probability and, for a detailed reading, a semester of real analysis), though it should be well within the background of upper-division engineering students. It is for this latter group that my book is written. As a mathematician, most of my friends at the university are engineers and physicists, virtually all of whom took micro- and macroeconomics as part of their school's core curriculum. They did not like it. It did not make sense and the mathematics was not being used to deduce theorems from axioms but only to illustrate conclusions that had already been agreed upon. Also, functions were never given explicitly so that mathematicians could deal with them. f(x)'s floated through economics texts without ever being specified and, one suspected, the variables and the relations between them were actually undefinable. While recognizing many of the faults in mainstream economics, my colleagues in the hard sciences were more concerned with studies in their majors and did not take time to develop a complete theory of economics. Without a theory of their own, it was frustrating for them to criticize a science which made no attempt at axiomatization or which only paid lip-service to some "axiom" which was never subsequently used to prove anything. One cannot find fault with a science's logic when no logic exists. For them, economics proved to be a very slippery target and few of its critics persisted. Nevertheless, it is a mistake to dismiss their criticism offhand. When beginners unanimously dislike a subject, experts should take note, for it is more likely that the subject is anti-intuitive than that every beginner lacks insight. Many beginners are leaders in their own field, perhaps even in a hard science.

Professional economists purposefully obfuscate their work. That way, only their supporters are able to decipher what the economist is advocating, usually more government intervention. Meanwhile the economist's opponents are failing to emphatically denounce him because they cannot find any quotations in his work that are written clearly enough to let anyone know what he is talking about. I am not a professional economist and can do nothing better for myself than to speak clearly to everyone who will listen. The greatest obstacle to promoting this book is the large number of Americans who gag at the first mention of economics, because they remember the horrible experience they had with it in college. It is not by obfuscating that I will reach these people.

The only virtue that an engineering student will readily grant an economics treatise is brevity, and I have tried to be as concise as possible. I am assuming that readers can motivate themselves. If they cannot, this book would not be for them no matter how many illustrations it contained, and if they can, they will appreciate how short it is. It is my hope that students in the hard sciences will find time to read my treatise. The introduction provides a simplified exposition of my theory and gives the reader an indication of the level of mathematics that will be employed. The first three chapters provide a foundation for economics and may be skipped by the less philosophically inclined. They do not contain any mathematics. Chapter Four treats much of the same material as the introduction but more thoroughly and without the simplifying axiom which allows the introduction to omit references to real analysis. Chapter Five does not require a mathematical background and is concerned primarily with monetary problems. Appendix C discusses mainstream macroeconomics. While the material in Appendix C has nothing to do with the main text, it is included here because I feel that it is important that I demonstrate a familiarity with mainstream theory before inventing my own. Mainstream macroeconomics is really very simple. It is just a matter of solving three equations in three unknowns. Unfortunately, because conventional textbooks can only print two-dimensional graphs, they spend semester after semester talking around their problem and never just say what the solution is. Also, they tend to get their units of measurement confused. I straighten out these problems and present the whole of their theory in one concise essay so that it can be understood and compared to my own.

Readers may visit my website at ***www.axiomaticeconomics.com***

✓ Download the **Simplified Exposition** of this theory.

✓ Download the BASIC program listed in Appendix B.

✓ Read reviews of **Axiomatic Theory of Economics**.

✓ Join our discussion group. The appendices discuss some unanswered questions or you may propose and prove theorems of your own. If any part of this work seems opaque, ask for help. I promise to make it just as confusing, only in a different way.

INTRODUCTION

I

I assert that one is capable of determining which of any two phenomena or sets of phenomena conform to a definition at a higher place on one's value scale than the other. If one fails to determine which of the two is higher, it can only mean that they are equal. In other words, one's value scale is a total (linear) ordering of phenomena. This is the first of three axioms which the reader is asked to accept. The plausibility of this axiom is derived mainly from analogy with the other dimensions (space and time), which are also totally ordered. A total ordering is included in the assertion of Absolute Geometry that every line has a coordinate system.

Because of this axiom, for every definition on one's value scale to which phenomena might conform, there stands beside it the number of units of money to which one is indifferent as to which one received. This supposition demands only that money be infinitely divisible, which it is for all practical purposes. As there are an infinity of distinct points on one's value scale, however, it cannot be expected that one is conscious of them all. In fact, one does not need to know exactly what one's point of indifference is to conduct many transactions.

The graph of the distribution of points of indifference, c(m), can be pictured as an aerial view of the people who value a phenomenon assembled along a line marked "money", where they are asked to stand by the number of monetary units that are equal to a unit of that phenomenon. If more than one person has the same valuation, they stand behind the corresponding number. The stock of that phenomenon naturally tends toward the high end, as anyone who possesses a unit of it who sees his neighbor to the right without one will sell it to him. Only use-value and expected exchange-value in other markets not represented on this graph are counted because, though one may value a phenomenon greatly in anticipation of exchanging it at a high price, if one fails to get that price, one has to lower one's asking price until it eventually equals the value of keeping that phenomenon for one's personal use. While money has very little use-value, it does have expected exchange-value in other markets not represented on this graph, and it is with this in mind that people withhold their money from this market if the price rises too high. The expected exchange-value of money is historically derived from its use-value. If it were a function of today's prices, we would have a contradiction because we are now deriving today's prices from the demand distribution, c(m), which includes expected exchange-value.

If a phenomenon has a steeply-diminishing utility for most people (after acquiring one unit, the importance of the next is very low because one easily becomes sated), most people are only represented once and c(m) is very close to $c_0(m)$, the distribution of people's point of indifference for their first unit. If there is a gradually-diminishing utility among people, many come back again and again before they become sated, each time with a lower point of indifference, and consequently the low end of c(m) rises. See figure 1. $R = \int_0^\infty c(m)dm$ is the requirement for a phenomenon by a population. Because stock is limited, however, only those with the highest use-value of it relative to their value of money possess any of the phenomenon. The price is less than the point of indifference of the last person who possesses a unit of the phenomenon or he would sell it, and it is greater than the point of indifference of the first excluded individual or he would buy. These two points of indifference are the marginal pair which determine the upper and lower limit of price, between which is the zone of indeterminacy. The formula relating price and stock to the demand distribution, c(m), is $S(m) = \int_{m}^{\infty} c(t)\,dt$ with S(m) the stock, m the price, and c(t) (t is a dummy variable for the integration) the distribution of points of indifference between the use-value of a unit of a phenomenon and t units of money. Of course, the expression above does not have any meaning until it is proven that stock converges. It will be used informally, however, until the convergence of stock is proven.

Figure 1

The Demand Distribution, c(m)

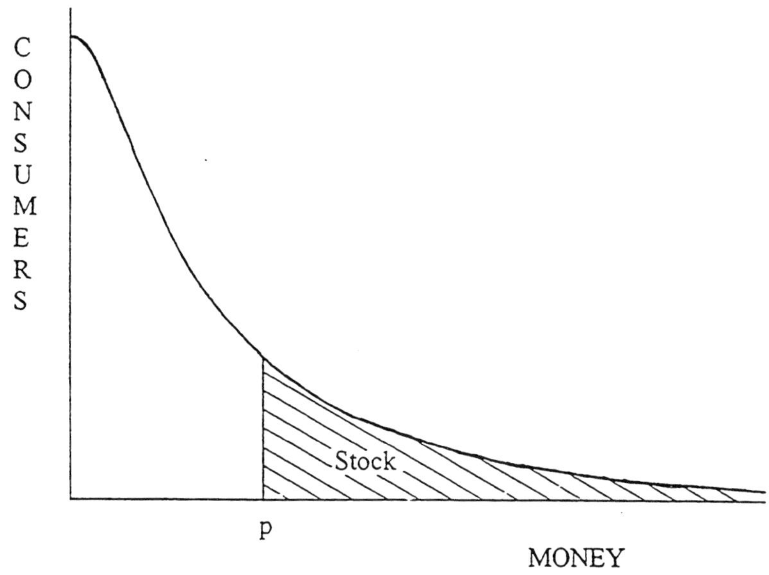

Both the people traditionally labeled "consumers" and those labeled "producers" appear in the demand distribution. The conceptual separation of consumers and producers is a great mistake of mainstream economics. They are all just people, each with a bit of the stock, and they are all prepared to sell if the price is above a certain point and buy if the price is below that point. The only thing that distinguishes people from one another is their point of indifference. This has little to do with who produced different bits of the stock, the event of production having occurred in the forgotten past. When economists draw one curve called "supply" and another called "demand", they are implying that the two are independent, for one cannot solve two simultaneous equations for two variables if the two equations are just versions of the same relation. Their dependence is well known at the macro level, but I assert that supply and demand are not independent at the micro level either. It is a mistake to inquire whether I support Say's assertion that "supply creates its own demand" or Keynes' assertion that "demand creates its own supply"; **Axiomatic Theory of Economics** is detached from that debate. I anticipate that the greatest block to the understanding of my theory will be people trying to interpret it in terms of supply and demand. I do not believe in supply and demand. I believe in the demand distribution, which is a mapping between price and stock. Supply has no place at all in **Axiomatic Theory of Economics**. My theory is not even divided into "micro" and "macro" sections. These terms were invented by mainstream economists when it became necessary to paste Keynes' theory over the top of Marshall's theory. They are clearly incompatible and their association in modern textbooks is entirely due to the bookbinder, not the economist.

By what criterion does mainstream economics distinguish people represented on a supply curve from those represented on the associated demand curve? This is a particularly pressing question for people dealing in narcotics because the penalties are so much greater for being on one curve than the other. But, if one visits a neighborhood where such trade takes place, any of the people one encounters would sell if the price were right and would buy if offered a bargain. There is really only one relation and it is called the demand distribution. Since there are two variables, price and stock, this (single) relation can provide a mapping from one variable to the other but cannot fix them. However, later in this introduction, existence and uniqueness proofs are given for a point toward which price and stock tend. Thereafter, it will be assumed that they are fixed at that point, called saturation.

The method of mainstream economics really has a third variable which is never mentioned and that is the time unit for supply and demand. It is well known that elasticity is a function of this time unit and, if this is true, one calculates a different price depending on whether one speaks of weekly or monthly supply and demand. This is an inconsistency since there can only be one price and it is not dependent on the caprice of an economist when he decides how often to conduct his surveys. This is a point that is glossed over in mainstream texts. A detailed discussion of the time unit chosen for supply and demand is never given and many texts neglect to mention the need for choosing one at all. Yet in their chapter on elasticity, every textbook lists time as a factor, sometimes as the most important factor.

Mainstream economists have two variables, price and quantity per unit of (some usually unspecified) time, and two equations, supply and demand. For this to work at all, the equations must be independent, which means that each individual must be either a buyer or a seller. The economist's decision to put people on one curve or the other cannot depend on the price that they would buy or sell at because both equations are defined for all prices. (Price is one of the independent variables.) So what is the economist's decision based on? Ask him repeatedly until he admits that there is really only one distribution. Also, press him to acknowledge that the demand distribution independently exists at each instant of time. Supply and demand curves are different depending on the time unit chosen. Mainstream economists provide no proof that their predicted prices are independent of their choice of time unit. For example, will thirteen predicted weekly quantities be the same as three predicted monthly quantities?

A large part of the problem with supply and demand is that it is used descriptively, but called predictive. It is easy to predict the past. Economists just observe the quantity produced one month and what it sold for and they put a little × over that spot. Then, by pure conjecture, they draw four tails on their × to fill their graph paper. Supply and demand has never been used predictively, not even to make bad predictions. × marks the spot is a purely descriptive technique. Since they are using the 20-20 vision of hindsight, they can do this for three months in a row and, to nobody's surprise, the sum of the quantities is the quarterly quantity. In the real world, price is constant for years at a time but, for most companies, their weekly and monthly sales figures swing wildly and unpredictably, sometimes by several fold from one month to the next. Mainstream economists have no explanation for this, which they should since their theory is called supply and demand and the horizontal axis of their graph is labeled weekly (or monthly) quantity. When I have been asked to help predict sales, I have told them that price is related to stock, not supply, and that they should stop watching their sales chart so ardently. At most companies, there is someone in accounting who feeds sales figures to the employees so that they can predict layoffs. They know that every dip in sales will send hundreds of them to the unemployment office, and that every rise will have their bosses clapping each other on the back and extolling their brilliant and farsighted management. They also know that nobody can predict sales. Supply never means anything in economics, though sometimes (for non-durable phenomena) it can pass for stock.

It is well known that mainstream economics is in trouble. Nobody in the hard sciences respects economists and even within their own ranks, a number of books and articles have appeared questioning why economics is not yet a science. There is considerable debate among economists about methodology, what it takes to qualify as a science, and what distinguishes economics from other fields. Implicit throughout is the understanding that mainstream economics does not work. To qualify as a science, economics must be axiomatic. But one must address price and stock; supply and demand does not work. Also, to deduce mathematical expressions from axioms, the axioms must be of a mathematical nature and they must specify actual functions from which equations can be derived. Fortunately, however, there is nothing fundamental about economics that prevents it from being made into a science just as physics was made into a science by Newton and mathematics by Euclid. This is what I propose to do.

II

There is an upper bound to one's value of any stock of a phenomenon which will be denoted M. This includes one's need for saving phenomena for future use. Total utility is the marginal utility of a phenomenon when the unit is defined as the entire quantity possessed. It increases from zero up to its maximum point, M, as one's stock increases. Hence, total utility is a cumulative distribution function and marginal utility is the associated probability density function (after normalization), denoted U(s) and u(s), respectively. Since the utility of a given stock is measured by the quantity of money which stands beside it on one's value scale, U(s) is a mapping from the stock of a phenomenon one possesses to the money one associates with that stock. u(s) is its first derivative. u(s) must be negative monotonic because utility diminishes as one adds units to one's stock. The integral of u(s) must also converge, that is, $\int_0^\infty \mathrm{u}(s)ds < \infty$. This is because marginal utility is the probability density function of a cumulative distribution. Nothing else is known about u(s) and this is the first parameter (and the only function) used to distinguish phenomena from one another. Its characteristics must be regarded as an axiom. Later, two more parameters (both from $\Re^+$) will be introduced which will be sufficient to completely describe every phenomenon. A graph of U(s) and u(s) is given in figure 2.

As will be shown shortly, we are only concerned with the ratio $\frac{\mathrm{u}(0)}{\mathrm{u}(r)}$ for non-negative integers, r. This ratio is invariant under a rescaling of the vertical axis, so u(s) can be normalized by setting the upper bound on the distribution function, M, to unity. This makes u(s) a true probability density function as the total area under it is unity.

Figure 2

The Relation Between Total and Marginal Utility

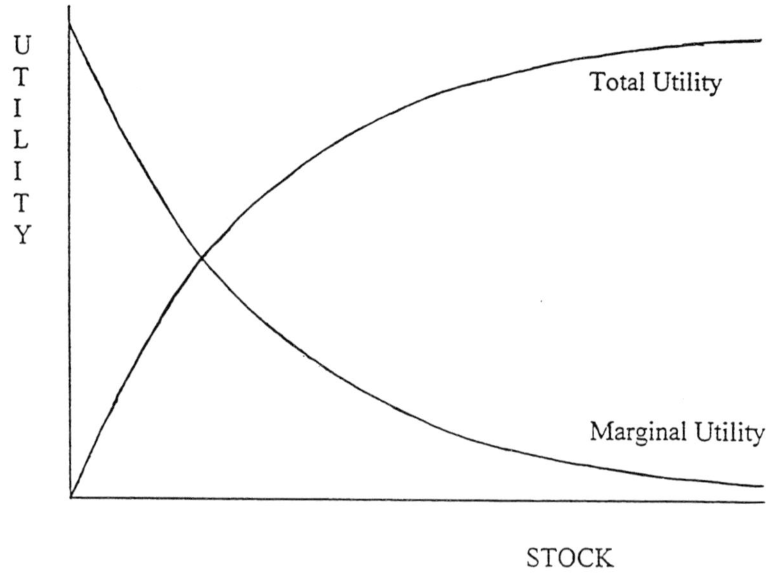

It should be noted here that the requirement that u(*s*) be negative monotonic does not imply that firms must be small, which is clearly not true because there are many large and successful corporations. Economists have used the term "marginal (or diminishing) utility" to denote both the first derivative of one's total utility for some phenomenon and the assertion that firms receive less and less return on their investments as they grow bigger. Capital, like all phenomena, has diminishing utility because one quickly becomes sated on it. However, like most things that one temporarily sates oneself on, one is ready for more the next day and the day after that. Thus, while a firm cannot immediately make use of all the capital it might consider buying, it can start with a small capital project and use the profits from that to train the managers and laborers that will make an expansion feasible. In this way, firms can become global in scale without ever contradicting the assertion that u(*s*) is negative monotonic for capital. The large corporation embarking on another great expansion may have started out as a small mom-and-pop outfit, but it is not that little company anymore and it has a (very) different utility function now. Since **Axiomatic Theory of Economics** is about stock, not supply, the relative sizes of the firms supplying a phenomenon is of no concern.

I assert that the distribution of people's points of indifference for their first unit of a phenomenon relative to money, $c_0(m)$, is lognormal; that is, the natural logarithm of the number of people who are indifferent at a particular price, *m*, is cumulatively (normally) distributed. The cumulative distribution is applicable to a variable that is subject to a process of change such that, at each step, a random quantity is added to the accumulated value of that variable. By the Central Limit Theorem, the distribution of the sum of a large number of independent, identically-distributed random variables (from an unspecified distribution with a finite mean and a non-zero, finite variance) is approximately normal. $c_0(m)$, however, does not accumulate, rather it is analogous to the growth of the value of money through history: It conforms to the characteristics of proportionate effect. After the j'th day of a person's life, the change in the number of monetary units to which he is indifferent, relative to the first unit of a phenomenon, is a proportion of his indifference point the day before. That anthropometric variables (height, size of organs, tolerance to drugs, etc.) conform to the characteristics of proportionate effect is well established in the literature.

Theorem 1 (Law of Proportionate Effect): Phenomena which conform to the characteristics of proportionate effect are lognormally distributed.

Proof:

$m_j - m_{j-1} = \varepsilon_j m_{j-1}$	The difference between each step and the last one is the last one multiplied by a random quantity.

$$\frac{m_j - m_{j-1}}{m_{j-1}} = \varepsilon_j$$

Divide through by m_{j-1} to get ε_j, the change in m relative to its previous value, m_{j-1}.

$$\sum_{j=1}^{n} \frac{m_j - m_{j-1}}{m_{j-1}} = \sum_{j=1}^{n} \varepsilon_j$$

Find the sum of all ε_j from the initiation of the process to its termination after n steps.

$$\int_{m_0}^{m_n} \frac{dm}{m} = \sum_{j=1}^{n} \varepsilon_j$$

If each step is small, $m_j - m_{j-1}$ can be approximated by dm.

$$\ln|m_n| - \ln|m_0| = \sum_{j=1}^{n} \varepsilon_j$$

Integrate from m_0 to m_n.

$$\ln|m_n| = \ln|m_0| + \varepsilon_1 + \ldots + \varepsilon_n$$

Solve for $\ln(m_n)$.

As can be seen from the last step, the natural logarithm of one's indifference point after the n'th day is a constant (the logarithm of its initial quantity) with a large number of random and identically-distributed quantities accumulated onto it. Hence, after having lived through n days and having seen their point of indifference change by a small proportion each day, consumers of their first unit are normally distributed with regard to the variable $\ln(m)$ and, hence, are lognormally distributed with regard to the variable m. ■

The absolute value operation may be dropped, since we are only interested in positive prices.

That first-unit demand conforms to the characteristics of proportionate effect must be regarded as an axiom. A plausibility argument is provided here. Let $m_j = \phi(m_{j-1})$ with m_j the number of monetary units to which one is indifferent relative to the first unit of a phenomenon on the j'th day of that person's life. We want to show that $\phi(m_{j-1}) = (1+\varepsilon_j)m_{j-1}$. Consider a man who wants to take out a loan at interest. He must think he will have more money in the future than he does now. (More money holdings, not necessarily more wealth.) If he does, the value of individual monetary units will tend to decrease over time relative to other phenomena; that is, ϕ is a positive function when averaged over all phenomena. To determine how much interest he is willing to pay, the man must specify this average ϕ. For him to calculate the interest owed per unit of time as a percentage of the principle is equivalent to specifying $\phi(m_{j-1}) = (1+\varepsilon)m_{j-1}$ with $\varepsilon > 0$ fixed. Fixing ε is a special case of ε_j being a random variable. Here, the probability density function is unity at ε and zero elsewhere. Thus, the axiom that first-unit demand conforms to the characteristics of proportionate effect is a generalization of calculating interest as a percentage of the amount owed. In fact, this is how people have calculated interest throughout recorded history, although economics having always been a soft science, they never asked for proof. Perhaps the value of money decays harmonically over time or in another way besides exponentially? This question is addressed in the main text, but for now let us proceed to investigate the consequences of people's points of indifference for their first unit of each phenomenon being lognormally distributed. I believe that this axiom is on solid intuitive ground and will not be criticized. Even if it is, it is unlikely that critics will succeed in convincing the banking industry to calculate interest with a different formula, so the weight of tradition will continue to support my choice of the lognormal distribution for first-unit demand.

Before continuing, let us explicitly state our three axioms:

1) One's value scale is totally (linearly) ordered:

i)	Transitive;	$p \leq q$ and $q \leq r$ imply $p \leq r$
ii)	Reflexive;	$p \leq p$
iii)	Antisymetric;	$p \leq q$ and $q \leq p$ imply $p = q$
iv)	Total;	$p \leq q$ or $q \leq p$

2) Marginal (diminishing) utility, $u(s)$, is such that:

- **i)** It is independent of first-unit demand.
- **ii)** It is negative monotonic; that is, $u'(s) < 0$.
- **iii)** The integral of $u(s)$ from zero to infinity is finite.

3) First-unit demand conforms to proportionate effect:

i) Value changes each day by a proportion (called $1+\varepsilon_j$, with j denoting the day), of the previous day's value.

ii) In the long run, the ε_j's may be considered random as they are not directly related to each other nor are they uniquely a function of value.

iii) The ε_j's are taken from an unspecified distribution with a finite mean and a non-zero, finite variance.

$\ln(m)$ is linearly transformed by $\frac{\ln(m)-\mu}{\sigma}$. The location parameter, μ (mean), quantifies the importance of a phenomenon relative to money and the scale parameter, σ (standard deviation), quantifies the difficulty of substituting other phenomena for the one in question. Easily-substituted phenomena have very little probability in the tail of their demand distribution; only the eccentric purchase a phenomenon at a high price when there are cheaper substitutes available. As substitution becomes more difficult, people must purchase the phenomenon even at high prices, and their distribution is less skewed. Both μ and σ must be positive. With $u(s)$, μ and σ describe all phenomena. Thus, every phenomenon is associated with a point in $u(s),\mu,\sigma$ space where $u(s)$ is a negative-monotonic probability density function on $\Re^+$ and μ and σ are both from $\Re^+$. For the purpose of economics, nothing else distinguishes one phenomenon from another.

The equation for the distribution of first-unit demand is $c_0(m) = \frac{e^{-\frac{1}{2}\left(\frac{\ln(m)-\mu}{\sigma}\right)^2}}{\sigma m}$. This is the equation of the lognormal distribution, $e^{-\frac{\ln^2(m)}{2}}$, multiplied by the derivative of the linear transformation which is substituted for $\ln(m)$. It need not be divided by its total area, $\sqrt{2\pi}$, since it will not be used as a probability density function.

The number of people with a point of indifference at a particular price, m, for their first unit is $c_0(m)$. Whoever's point of indifference for his first unit is $\frac{u(0)}{u(1)}$ times greater than that price values his second unit equivalent to price *m*. Whoever's point of indifference for his first unit is $\frac{u(0)}{u(2)}$ times greater than that price values his third unit equivalent to price *m*, and so on. To find c(*m*), all the people with a point of indifference at *m* are summed up, whether it is their first purchase or a later purchase. Recall the analogy of the demand distribution being an aerial view of the people who value a phenomenon assembled along a line marked "money", where they are asked to stand by the number of monetary units that are equal to a unit of that phenomenon. Now consider a person who wishes to possess more than one unit of the phenomenon; each of his agents appears behind a different point on the money line. If he himself appears in the column assembled behind *m* monetary units, the first person he sends to get another unit is directed to the column behind $\frac{mu(1)}{u(0)}$ monetary units. His next agent is in the column behind $\frac{mu(2)}{u(0)}$ monetary units, and so on. Hence, we have the following formula for the demand distribution which, unfortunately, is impossible to integrate in closed form, even with u(*s*) fixed.

$$c(m) = \sum_{r=0}^{\infty} c_0(x) \qquad \text{with } x = \frac{mu(0)}{u(r)}$$

$c_0(m)$ can be thought of as the 0'th partial sum of c(*m*) and, in general, $c_n(m)$ denotes the n'th partial sum of c(*m*). Thus,

$$c_n(m) = \sum_{r=0}^{n} c_0(x) \qquad \text{with } x = \frac{mu(0)}{u(r)}$$

Most of the real analysis in **Axiomatic Theory of Economics** stems from the infinite summation, c(*m*). To simplify the proofs in this introduction, the second axiom is replaced with the assertion that people never need more than one of anything at a time. This assumption is neither accurate nor necessary, as all of the results of my theory can be (and are) proven in their full generality. However, some economists do not have the mathematical background necessary to read my book, so, for expository purposes, simplified proofs are provided here. Also, before the theory becomes accepted, it will receive cursory reviews, perhaps at the end of courses on mainstream economics. In this case, if a professor is sympathetic to my theory, he may wish to prove some of its assertions, but he will not have time to prove them in their full generality. As long as he mentions that the complete proofs do exist, his students can get the essence of my theory from the simplified proofs. The important thing for them to understand is that this theory is deduced from axioms. So, for the remainder of this introduction, all of the theorems will be proven using the 0'th partial sum, $c_0(m)$, rather than $c(m)$. When $S(m)$ appears in a proof, it will refer to $S(m) = \int_m^\infty c_0(t)dt$. $f(\mu,m)$, which will be defined later, will also be defined in terms of $c_0(m)$ rather than $c(m)$.

III

Theorems are numbered analogous to those in the main text.

Theorem 4: $\lim_{m \to 0^+} c_0(m) = 0$

Proof: $c_0(m) > 0$ for all $m > 0$. Thus, by the Squeezing Theorem, if $c_0(m)$ is less than some function for all $m > 0$ and that function is continuous and equals zero at zero, then $\lim_{m \to 0^+} c_0(m) = 0$. Consider h$m$ with h a finite constant. Since hm vanishes at zero, it is sufficient to show that $hm > \dfrac{e^{-\frac{1}{2}\left(\frac{\ln(m)-\mu}{\sigma}\right)^2}}{\sigma m}$ for all $m > 0$. By making the substitution $y = \ln(m)$, this is equivalent to $y^2 + (4\sigma^2 - 2\mu)y + 2\sigma^2\ln(\sigma h) + \mu^2 > 0$ for all real y. By the Quadratic Theorem, this is true for $\dfrac{e^{2(\sigma^2-\mu)}}{\sigma} < h < \infty$. Thus, the demand distribution is equal to zero at zero. ■

Alternate proof: Make the substitution $y = \ln(m)$ so

$$\lim_{m \to 0^+} c_0(m) = \lim_{y \to -\infty} \frac{e^{-\frac{1}{2}\left(\frac{y-\mu}{\sigma}\right)^2}}{\sigma e^y}$$

$$= \frac{1}{\sigma} \lim_{y \to -\infty} e^{-\frac{(y-\mu)^2 + 2\sigma^2 y}{2\sigma^2}}$$

$$= \frac{1}{\sigma} \lim_{y \to -\infty} e^{-\frac{(y-\mu+\sigma^2)^2 - \sigma^2(\sigma^2 - 2\mu)}{2\sigma^2}}$$

$$= 0 \ \blacksquare$$

The former was chosen as the main proof because the Squeezing Theorem and the Quadratic Theorem can be visualized and are (hopefully) more intuitive than a purely algebraic proof.

Theorem 7: Stock is finite.

Proof: Make the substitutions $y = \frac{\ln(t)-\mu}{\sigma}$ and $dy = \frac{dt}{\sigma t}$ so

$$S(m) = \int_z^{\infty} e^{-\frac{y^2}{2}} dy \qquad \text{with } z = \frac{\ln(m)-\mu}{\sigma}$$

The integral is the standard normal distribution, which is tabulated as $\alpha(z) = 1 - \Phi(z)$ in the back of any statistics text, though multiplied by the constant $\frac{1}{\sqrt{2\pi}}$ so that the total area under the integrand is unity, a step which is omitted here since the integrand is not being used as a probability density function. However, since this integral never exceeds $\sqrt{2\pi}$, we have the following inequality: $S(m) < \sqrt{2\pi}$. ■

Aggregate utility is defined as price multiplied by stock. This is because money is the measure of utility and everyone who possesses a unit of stock values it only as highly as its replacement cost, for that is all that one risks. Stock and price, however, are inversely related, so increasing one or the other does not necessarily increase aggregate utility. Aggregate utility being the common goal of people dealing in a phenomenon, they are interested in maximizing it. As stock increases, aggregate utility also increases up to saturation, where any further increases in stock reduce aggregate utility by driving the price down.

That part of the demand distribution to the right of saturation (the high end), where increases in stock increase aggregate utility, is unsaturated and that part to the left (the low end) is saturated. At a constant stock, there is a zone of indeterminacy between the marginal pair within which the price may fluctuate. Such fluctuations appear to be of a saturated market whether the stock has reached saturation or not. Most markets are large enough, however, that the zone of indeterminacy is too narrow to be of practical concern.

Because the actions appropriate in an unsaturated market (increasing stock) are not those appropriate in a saturated market (decreasing stock), it is important to determine the point of saturation. Aggregate utility, $m\mathrm{S}(m)$, is at a relative maxima where its first derivative, $\mathrm{S}(m) - m\mathrm{c}(m)$, equals zero. Thus, saturation is a price and stock such that $\mathrm{S}(m) = m\mathrm{c}(m)$. Graphically, $\mathrm{S}(m)$ is represented by the area between the horizontal axis and the graph of the demand distribution from m to ∞. $m\mathrm{c}(m)$ is represented by the area of the rectangle formed by the two axes and horizontal and vertical lines extending from the point $m,\mathrm{c}(m)$.

Theorem 10 (existence): The absolute maximum of aggregate utility is at a finite critical point.

Proof: By Theorem 4, the limit of $\mathrm{c}_0(m)$ at zero is zero. Thus, stock is finite even if it is free, and aggregate utility goes to zero as price approaches zero. Since aggregate utility is always positive, it is sufficient to show that it also goes to zero as price approaches infinity to prove the existence of a relative maxima. One makes the substitutions $y = \frac{\ln(t)-\mu}{\sigma}$ and $dy = \frac{dt}{\sigma t}$ so

$$0 < m\mathrm{S}(m) = m \int_{\frac{\ln(m)-\mu}{\sigma}}^{\infty} \mathrm{e}^{-\frac{y^2}{2}}\, dy$$

$$\leq m \int_{\frac{\ln(m)-\mu}{\sigma}}^{\infty} y e^{-\frac{y^2}{2}} dy \qquad \text{if } m \geq e^{\mu+\sigma}$$

$$= -m \left[e^{-\frac{y^2}{2}} \right]_{\frac{\ln(m)-\mu}{\sigma}}^{\infty}$$

$$= m e^{-\frac{1}{2}\left(\frac{\ln(m)-\mu}{\sigma}\right)^2}$$

$$= m e^{-\frac{\ln^2(m)-2\mu\ln(m)+\mu^2}{2\sigma^2}}$$

$$= \frac{B m e^{-\frac{\ln^2(m)}{2\sigma^2}}}{m^{-\frac{\mu}{\sigma^2}}} \qquad \text{with } B = e^{-\frac{\mu^2}{2\sigma^2}}$$

$$= \frac{Bm}{m^{\frac{\ln(m)-2\mu}{2\sigma^2}}}$$

$$\leq \frac{B}{m} \qquad \text{if } m \geq e^{2(\mu+2\sigma^2)}$$

B is a constant, so $\lim_{m\to\infty} B/m = 0$ and, by the Squeezing Theorem, $\lim_{m\to\infty} mS(m) = 0$. Thus, there exists a finite price where aggregate utility is at a maximum. ■

This only proves the existence of a relative maxima and identifies it with the absolute maximum. There may be more than one relative maxima, in which case the largest of them is the absolute maximum. However, by the following proof there is only one relative maxima and it is the absolute maximum of aggregate utility. This justifies the use of the word "the" when referring to the saturation point.

Theorem 11 (uniqueness): Aggregate utility has only one relative maxima.

Proof: Because aggregate utility is always positive and it approaches zero at both ends of its domain, $(0,\infty)$, there is either a single relative maxima, or relative maximas and minimas alternate with the largest and smallest being relative maximas. The second derivative of aggregate utility is $c_0(m)\left(\frac{\ln(m)-\mu}{\sigma^2} - 1\right)$. It is positive at relative minimas and negative at relative maximas. Therefore, if there is more than one relative maxima, there are two disjoint intervals in $(0,\infty)$ where the second derivative is negative and they are separated by an interval where the second derivative is positive.

We wish to show where the second derivative is strictly negative. $c_0(m) > 0$ for all m, so we only have to examine $\frac{\ln(m)-\mu}{\sigma^2} - 1$. This is negative for all $0 < m < e^{\mu+\sigma^2}$ and positive for all $m > e^{\mu+\sigma^2}$. Recalling that relative maximas and minimas alternate with the largest and smallest being relative maximas, there can only be one price such that $S(m) = mc_0(m)$ and it is a relative maxima. ■

It is an easy corollary that the saturation price is less than $e^{\mu+\sigma^2}$.

μ and σ change over time for a variety of reasons, each change necessitating a recalculation of the saturation point. It is the business of entrepreneurs to anticipate these changes and to adjust stocks accordingly. While most shifts in a demand distribution are of only local concern, one is of particular interest to economics. If some of the people represented by the demand distribution for a phenomenon receive money from the government, how does the saturation point change? Whether these people receive a grant, a low interest loan, or are doing contract work for the government, they are more liquid than they want to be. Knowing the negative effect of a loose monetary policy on the value of money, they are not going to hoard it. Relative to money, the importance of phenomena has increased. How are prices and stocks affected and which adjusts more dramatically to the increase in μ?

It is an old adage that people get more out of something the more they put into it and, money being the measure of utility, one expects increases in the importance of a phenomenon relative to money to increase the phenomenon's price in proportion to the price that it has already attained. Mathematically, $p = p_0e^{\mu}$, with p_0 the price at saturation with no importance relative to money and p the price such that $f(\mu,m) = S(\mu,m) - mc(\mu,m) = 0$. Notice that p is the particular price which satisfies the condition $f(\mu,m) = 0$ while m denotes an arbitrary price. Variables included in the functional notation are allowed to vary while others which appear in a function but are not listed in the parenthesis of the function are assumed to be constant. Here, we are discussing changes in both price and importance where before only price was allowed to vary.

Theorem 12: The price at saturation increases exponentially in response to an increase in the importance of a phenomenon relative to money; that is, $\dfrac{dp}{d\mu} = p$.

Proof: $f(\mu,m) = S(\mu,m) - mc_0(\mu,m) = 0$ implicitly defines a level set in the μ,m plane. Let that level set be parametized by $[\ \mu(t)\ \ m(t)\ ]$. By the chain rule, the derivative of $f(\mu,m) = 0$ is $\dfrac{\partial f}{\partial \mu}\dfrac{d\mu}{dt} + \dfrac{\partial f}{\partial m}\dfrac{dm}{dt} = 0$ or $\begin{bmatrix} \dfrac{\partial f}{\partial \mu} & \dfrac{\partial f}{\partial m} \end{bmatrix} \begin{bmatrix} \dfrac{d\mu}{dt} & \dfrac{dm}{dt} \end{bmatrix} = 0$. The latter vector is the derivative (tangent) of the parametized level set, so $\left[\frac{\partial f}{\partial \mu}\ \frac{\partial f}{\partial m}\right]$ is perpendicular to the level set which passes through any μ,m where it is evaluated. From the definition of saturation, this is downward (toward smaller m), so a 90° counter-clockwise rotation of $\left[\frac{\partial f}{\partial \mu}\ \frac{\partial f}{\partial m}\right]$ is tangent to the level set of all μ,m combinations with $f(\mu,m)$ constant. Dividing its vertical component by its horizontal component gives the desired rate of change in price:

$$\frac{dm}{d\mu} = -\frac{\frac{\partial f}{\partial \mu}}{\frac{\partial f}{\partial m}} = \frac{m\frac{\partial f}{\partial m}}{\frac{\partial f}{\partial m}} = m \qquad \text{with } \frac{\partial f}{\partial m} = c_0(m)\left(\frac{\ln(m)-\mu}{\sigma} - 1\right)$$

This relation is true regarding the level set which passes through any point μ,m. Choosing only points along the level set $f(\mu,m) = 0$ (rather than another constant) yields $\frac{dp}{d\mu} = p$. ■

Notice that $f(m)$ in the above proof may be expressed as

$$f(m) = -\int_m^\infty f'(t)dt = \int_m^\infty c_0(t)\left(1 - \frac{\ln(t)-\mu}{\sigma^2}\right)dt$$

Also, the evaluation of $\frac{\partial f}{\partial \mu}$ requires an application of Leibnitz' Rule, justification of which is given in the main text. Incidentally, it does not matter that the rotation is counter-clockwise since a clockwise rotation also switches the components but negates $\frac{\partial f}{\partial \mu}$ instead of $\frac{\partial f}{\partial m}$. Because the sign comes out front after the division, it is immaterial which way $\left[\frac{\partial f}{\partial \mu}\ \frac{\partial f}{\partial m}\right]$ is rotated.

An alternative proof uses the chain rule to differentiate $f(\mu,g(\mu,m)) = 0$ with $p = g(\mu,m)$ to get

$$f_{\mu}(\mu,g(\mu,m)) + f_{m}(\mu,g(\mu,m))g_{\mu}(\mu,m) = 0$$

This equation is solved for $\frac{dp}{d\mu} = g_{\mu}(\mu,m)$. Notice that, by the uniqueness of saturation, $p = g(\mu)$ is a function; that is, a unique price is associated with every μ, though in general this is not required for $g_{\mu}(\mu,m)$ to be determined explicitly. In other words, not every $g_{\mu}(\mu,m)$ has an anti-derivative, $g(\mu)$. By the construction of $g_{\mu}(\mu,m)$, $g(\mu,m)$ is proven to be smooth and continuous, which is all that is required of it.

Until this proof, only one semester of calculus had been required of the reader. Theorems 12 and 13 are about functions of two variables, however, and are more difficult. Readers with only one semester of calculus may find the alternative proof of Theorem 12 easier than the main proof if they are familiar with implicit differentiation. However, many students who have been introduced to calculus of several variables readily grasp the concept of level sets because of their familiarity with contour maps. Thus, for $f:\Re^{1+1}\rightarrow\Re^{1}$, recourse to the tangent seems more intuitive than a purely algebraic proof and the former was chosen as the main proof. Readers with only one semester of calculus can obtain most of the mathematics they need by reading a textbook on multivariable calculus up to but not including Lagrange multipliers. This is generally considered the easy part of multivariable calculus and is the work of six or eight lecture hours. To read the main text (without the simplifying axiom of the introduction) also requires some knowledge of infinite series. Fortunately, the "hard" part of multivariable calculus (multiple integrals and vector fields) is never used anywhere in this book. **Axiomatic Theory of Economics** is similar to probability. Indeed, I see my book following in the tradition of Kolmogorov's **Foundations of Probability** more than in any work of an economist. People who have worked with probability distributions are encouraged to read the main text even if they are only vaguely familiar with multivariable calculus.

By Theorem 12, the price at saturation increases exponentially in response to an increase in the importance of a phenomenon relative to money. What about stock? Intuitively, one expects stock to remain constant since, effectively, all the government does by issuing money is to change the figures in which prices are quoted and that should not affect the stock of phenomena that people keep in existence. Most economists would agree that this is true in the long run but would argue that, because of the uneven diffusion of fresh issues of money, the stock of phenomena is temporarily affected. Money diffuses unevenly from a central bank and that is the principal motivation for issuing it (otherwise those close to a government would not profit from their connections), but I assert that this does not provide any incentive for the stock of phenomena to increase.

Theorem 13: The stock at saturation remains constant in response to an increase in the importance of a phenomenon relative to money; that is, $\frac{dS_p}{d\mu} = 0$.

Here, the subscript on stock denotes that it is the stock associated with the saturation price, p.

Proof: We are interested in the change in stock along the level set implicitly defined in the μ,m plane by the relation $f(\mu,m) = 0$. As noted in the preceding proof, the tangent to this curve is $[\ 1 \quad m\ ]$. Normalizing this vector and taking the inner-product with the derivative of $S(\mu,m)$ gives the desired rate of change in stock. Since we are interested in proving that this change is always zero, it is sufficient to show that the numerator is always zero and we may omit normalizing the directional vector. The inner product of this with the derivative of stock, $[\ mc_0(\mu,m) \quad -c_0(\mu,m)\ ]$, is zero. ■

Together, the two preceding theorems will be referred to as the Law of Price Adjustment. A graph of $p(\mu)$ and $S_p(\mu)$ is given in figure 3. Because Theorem 13 is a corollary of Theorem 12, the term "Law of Price Adjustment" is used to denote both theorems. From a practical point of view, however, the assertion that the stock of phenomena is unaffected by depreciating a currency is more important because, by definition, economics is concerned with the wealth of a nation. Of course, the wealth of an individual can always be increased at the expense of other people by printing and spending money, but theoretical economics (hopefully) addresses more lofty aims.

Figure 3

The Law of Price Adjustment

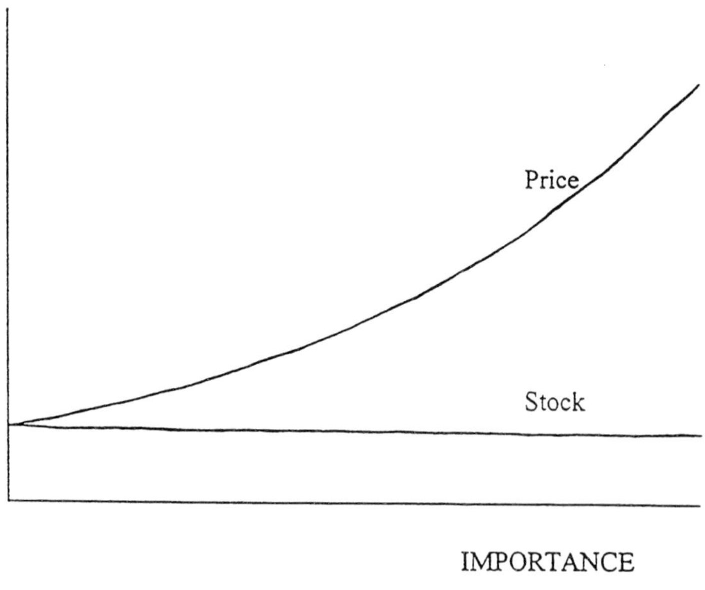

It is important that the Law of Price Adjustment does not place any restrictions on marginal utility, on the importance of a phenomenon relative to money, or on the difficulty of substituting other phenomena. Within my economic theory, these three characteristics are all that distinguish phenomena from one another; that is, phenomena with the same u(*s*), μ, and σ are isomorphic. Thus, it is impossible to argue that my theory is inapplicable in certain situations because it has been proven to apply to all possible situations; that is, it applies to phenomena at every point in $u(s),\mu,\sigma$ space. Since any mathematician will confirm the deduction of the Law of Price Adjustment from the three axioms, for an economist to accept or reject the Law of Price Adjustment is equivalent to his acceptance or rejection of the three axioms, respectively. Attempts to divert the argument away from the acceptance or rejection of the theory's axioms should be discouraged.

The implications of the Law of Price Adjustment should be obvious to anyone who has studied mainstream economics; stickiness of prices is the cornerstone of Keynesian Economics. Even for those who do not follow the mathematics, common sense alone is sufficient to refute the Keynesian premise. Considering that a government can print money for itself within a day's notice, if the adjustment process could not be done in equal time, the whole system of indirect exchange would have collapsed long ago. Prices can be changed with a word, but the stock of phenomena can only be changed after considerable toil. It is obvious which is adjusted and which left constant. The average level of prices is "sticky" because it takes time for money to diffuse through a community and if one is averaging all prices, it is some time before one notices a change in one's statistics. This average is also meaningless for the same reason. The effect of issuing money is to redistribute wealth to the people who receive the new money first and that is only possible because of the slow diffusion of money through an economy.

Having arrived at a position so fundamentally opposed to mainstream economics, it is important to realize exactly where we parted company. The difference is that my theory is concerned with the price and stock of phenomena while mainstream economics is concerned with the price and supply of phenomena. I assert that the stock of phenomena is more important than the supply because all of the decisions made regarding a phenomenon are based on its stock (how much of it is in existence), and not on how much of it happened to be produced in some arbitrary time period. Phenomena are the same whether they are produced in one time period or another. Most people do not know and none care what the supply of phenomena is, they are concerned with the stock; this week's or month's supply is only a small part of the available stock. Even if a factory is temporarily closed for a week or a month, the price of its product is hardly affected because the total amount of phenomena in existence is hardly affected. Yet during that week or month the supply is zero. Mainstream economics, which relates price to supply, is unable to explain why the price does not increase dramatically as inspection of the supply and demand curves predicts that it should.

Parking on campus has a price, so mainstream economists must believe that there is a supply, that is, an influx, of parking spaces. Yet none are being produced. Clearly, it is the stock, the absolute quantity of them, that determines price. Supply never means anything in economics, though sometimes (for non-durable phenomena) it can pass for stock. There are three principle mistakes of mainstream economics, but addressing supply and demand instead of price and stock is the most egregious. The other two are assuming that all short-term credit instruments function as money and believing that the average price level is a meaningful statistic and, hence, that prices are "sticky."

IV

The distinction between wealth and income is not just one of semantics. Most people define poverty to be low income, not a lack of wealth. It is unemployment figures, not asset figures, that are printed in the news media and it is national income that is maximized in IS-LM analysis. The logical conclusion of thinking in terms of a lack of jobs rather than a lack of wealth is either to put low-income people on the government's payroll or to train them to work. In fact, this is what most people have in mind when they think of a jobs program, and their conclusion is supported by mainstream economics. Anyone who has heard of the multiplier (it is introduced in first-semester macroeconomics) knows how mainstream economists think government spending affects national income.

However, I assert that welfare and make-work programs are not a solution to poverty because they do not create wealth. People just spend their subsidy on immediate consumption and, a month later, they are still in poverty and need another welfare payment. This is an easy application of the Law of Price Adjustment for, if such government spending created wealth, it would have to do so by increasing the stock of those phenomena which the recipients spend their subsidy on. But this contradicts Theorem 13, that the stock at saturation remains constant in response to an increase in the importance of a phenomenon relative to money. However, it may not be clear that job-training programs do not create wealth for, in a sense, job skills are as much a part of the capital stock as actual machines are. But job skills are not the most important part of the capital stock, primarily because they are so easy to acquire. Employers who ask for years of experience at relatively simple tasks do not accept training as a substitute. Their companies are expanding and they are hiring experienced personnel away from declining companies, which represents variance in an industry, not growth. Training people to operate certain machines does not work any more than welfare unless those machines are being manufactured;

all of the machines currently in existence have operators. Training programs are actually an insult to people's intelligence because they could learn to operate virtually any machine within a matter of days. They know that the reason they are unemployed is not a lack of skills but simply because there are not enough machines to go around. In general, there is not enough capital.

The act of creating new capital, manufacturing new machines, is called investment. A government has two methods to pay for its consumption: It can tax people or it can sell bonds. Taxes are just confiscation and are taken entirely out of private investment. Bonds that mature in the hands of private parties (not the central bank) compete with corporate bonds. Because the buyers hold them to maturity, they are interested only in receiving interest and would buy corporate bonds if the treasury did not offer them a higher interest rate. Thus, the two forms of fiscal policy, taxing and borrowing from the public, either confiscate or crowd out an equal amount of private investment. Bonds sold to the central bank (either directly or to private dealers who pass them on to the central bank before maturity) are paid for with a check that the central bank writes against itself. The treasury then deposits this check in a commercial bank, allowing commercial banks to buy government bonds amounting to several times the amount that the central bank bought. (The exact multiple is determined by the reserve requirement.) Thus, the two forms of monetary policy, selling bonds to the central bank and to commercial banks, are effectively just printing money and using it to buy capital away from private investors. Capital consumed by a government through fiscal and monetary policies was never idle but would otherwise have gone to the creation of wealth. Common sense alone is sufficient to verify such an obvious assertion.

The use of credit money (bills of debt) is an alternative to using money certificates and thus a negative influence on their value. It is sometimes thought that, if the issue of money certificates, which is also a negative influence because it increases their quantity, is used only to eliminate the first negative influence, the value of money remains stable. Thus, the argument continues, if a central bank refrains from issuing money certificates except to buy bills of debt which are then held until they mature, the quantity of fiduciary money in circulation is made directly proportional to the quantity of credit money removed from circulation, leaving the stock of money in the broader sense unchanged. This would work if it were not for the fact that the quantity of credit instruments removed from circulation (sold to a bank) is itself a function of the quantity of fiduciary money issued by the bank. When a person writes an IOU for phenomena received, it has a face value greater than the present price of that phenomenon or the seller would demand immediate payment. If the seller of that phenomenon is unable to wait for the IOU to mature, however, he can sell it to a bank at a discount of its face value, thus receiving only what he would if he demanded immediate payment for his phenomena. Because of his weakness and the bank's relative strength when it comes to the ability to wait for an investment to be realized, the bank collects interest on that IOU. This strength of a bank's, like anyone else's, is a function of its wealth. However, because a central bank has the ability to arbitrarily increase its wealth with fiduciary money, it has a disproportionate advantage. This allows it to buy credit instruments at very close to their face value even when they still have some time to mature. This practice greatly increases the demand for credit instruments instead of immediate payment because people know that they can sell them to a bank. Thus, credit instruments are sold to a bank that were never really a part of credit money as they were never exchanged from person to person, which would have decreased the need for money certificates, but were sold to the bank immediately upon being written. Existence alone is not a sufficient condition for credit instruments to be included in the stock of money; there must be an active secondary market for them in the community.

Thus, the institution of central banks is built on a deception. By assuming that all short-term credit instruments function as money, they can issue money certificates while claiming to leave the stock of money in the broader sense unchanged. If they operated like counterfeiters and just printed money and then went out into the world to spend it, they would receive little support from the people. Instead, they accomplish the same thing but in such a circuitous manner that they receive only confusion from the people. The treasury issues bonds which may be purchased by anybody but are mostly purchased by private dealers. Most of these are resold to the central bank. The check it writes is deposited at a commercial bank where it is highly valued because it counts in the reserve requirement that commercial banks must keep to back up their own checking accounts. Cash also counts, but the checks of other commercial banks do not. The reserve requirement is about 12.5% and, while the bank which receives a check from the central bank cannot itself write eight times that amount in checks, by a process explained in the main text, the amount of checks written by all commercial banks will increase by eight times the amount of the central bank's purchase of government bonds. The commercial banks can spend this new-found money on anything they want, though they are encouraged to spend it on government bonds. Without a vast and bloated bureaucracy to watch over them, however, it is difficult to prevent the banks (including savings and loans) from spending it on get-rich-quick schemes which, if successful, bring great profits to the bankers and, if unsuccessful, dump great losses on the taxpayers. The term "national debt" refers to the government bonds sold mostly to the central bank, some to commercial banks, and almost none to private savers. Since the central bank and the treasury are both branches of the government, they have effectively just printed some money and spent it. The word "debt" is used only to obscure the process and does not have any meaning in this context. One possible exception is countries with a large trade deficit. This implies that foreigners own some of their assets and, while they usually prefer real estate and businesses, they may take government bonds.

To attack the root cause of unemployment, the government must be prevented from wasting the capital of the nation. Piecemeal elimination of obvious boondoggles is a move in the right direction. However, because of the vested interest in each boondoggle, the system of pork-barrel politics is quite stable. It is more effective to reduce the government's revenue and leave Congress alone to spend what they get than it is to attempt to influence specific legislation. Revenue from taxes has a natural limit: People individually evade taxes that they consider to be unfair and collectively vote against legislators who support unreasonable tax bills. Tax revenues are responsive to genuine emergencies such as an invasion, however, so they may be considered the measure of how much government people want. To reduce the revenue of a government down to what people are willing to pay in taxes (or patriotically purchase in bonds), the central bank must be eliminated. In the United States, the central bank is the Federal Reserve. If it can be shown to be unstable, then decentralization of the banking industry is possible.

Decentralization is not the same thing as deregulation. The term "regulation" is meaningless without reference to the basic framework in which banks operate. A stable system can be governed by the usual laws against criminality that apply to all businesses, while an unstable system requires a vast regulatory bureaucracy and is still plagued with corruption. It is naive for people who dislike big government to advocate deregulation in the latter case, but it is also wrong to assume that the existence of a central bank is part of the regulations which attempt to prevent corruption. Central banks and regulatory bureaucracies are associated with one another, not because they both oppose an inherent instability in banking, but because the existence of a central bank creates an unstable system that requires constant policing.

The theory of economics presented in this book is divided into two chapters and they each have a point. The point of the first chapter is the Law of Price Adjustment. The point of the second chapter is that the purpose of monopolizing the right to issue money is to cheapen it. If every bank issued its own money, none of them could cheapen it for fear of losing reserves to competing banks and the system would be perfectly stable. A central bank does not provide stability because, having eliminated its competition, it need not fear losing reserves due to imprudent management. In fact, without massive regulation, it causes just the kind of corruption that was seen in the United States' recent, misguided attempt at deregulation. The purpose of the Federal Reserve is not now and never has been to create stability. Its purpose is to provide the government with a convenient method to siphon off the wealth of the nation. A decentralized system does not do this because, without a central bank standing ready to fill the treasury, it is unlikely that private banks would consider the government a good enough risk to grant it credit. At least they would not grant the government unlimited credit, which is what the Federal Reserve was designed to do.

Eliminating the Federal Reserve is more efficient than requiring that the government balance its budget. Balancing the budget is an accounting trick and to talk about a budget while a central bank exists is, quite frankly, missing the point. That would be like successfully besieging a city and then telling one's soldiers to be sure that they pay for anything they take from the people's homes and shops. The soldiers would think that their general was mad. The purpose of besieging a city is to loot it and the purpose of a central bank is to run deficits. And the national debt, lest anyone misunderstand, is all that the government has consumed without the consent of taxpayers.

Note to the reader: I have written a pamphlet entitled **Simplified Exposition of Axiomatic Theory of Economics**. The pamphlet is essentially a reprint of this introduction. It can be downloaded from ***www.axiomaticeconomics.com*** or you may order copies from the address below. Printing and mailing costs require a $5.00 payment. As long as my name appears on every page, the pamphlet may be copied as often as you wish. You may give the copies away or sell them to recoup your copying and distribution costs. Copies may not be bound or packaged with other material or altered in any way.

Victor Aguilar
7620 E. McKellips #424
Scottsdale, AZ 85257

www.axiomaticeconomics.com

PART ONE

FOUNDATIONS OF ECONOMICS

The sun of quality does not revolve around the subjects and objects of our existence. It does not just passively illuminate them. It is not subordinate to them in any way. It has created *them. They are subordinate to* it... *Now to take that which has caused us to create the world and include it in the world we have created, it is clearly impossible. That is why quality cannot be defined.*

Robert Pirsig
Zen and the Art of
Motorcycle Maintenance[1]

Physical concepts are free creations of the human mind, and are not, however it may seem, uniquely determined by the external world. In our endeavor to understand reality we are somewhat like a man trying to understand the mechanism of a closed watch. He sees the face and the moving hands, even hears the ticking, but he has no way of opening the case. If he is ingenious he may form some picture of a mechanism which could be responsible for all the things he observes, but he may never be quite sure his picture is the only one which could explain his observations. He will never be able to compare his picture with the real mechanism and he cannot even imagine the possibility of the meaning of such a comparison.

Albert Einstein
The Evolution of Physics[2]

CHAPTER I

AGNOSTICISM

How angry He got with us, this wrath-snorter, because we understood Him badly! But why did He not speak more clearly? And if it was the fault of our ears, why did He give us ears that heard Him badly? If there was mud in our ears - well, who put it there?

Frederick Nietzsche
Thus Spoke Zarathustra[3]

Section 1

1.1 What is good and what is not good?

This is the question of aesthetics and around it speculative philosophy has revolved since time immemorial. As one opens one's eyes, they are filled with a cascade of light, and when one reaches out one's hand, manifold forces press against it. Some of these are called good and some are called not good, but where is the distinction? Is it in the events themselves? No, for all that they contain are patterns of light, forces against one's touch, and vibrations against one's ear drums. Value is no more contained in the particles of light that strike one's retina or in the air molecules that are bounced against one's ear drum than location is contained in them. These events are only quanta of energy; value and location are supplied by one's mind. Time is also a product of the mind, for the one particle of light or molecule of air does not contain the memory of those before it. The hoard of sense stimuli striking one's nerve endings is not the picture one perceives. It is only a hoard of sense stimuli striking one's nerve endings. There are no objects, only events; from the first not a thing is. The process of ordering perceptions into good and bad, here and there, now and then, is entirely *a priori*. One's mind is not a *tabula rasa* on which one's sensations impinge and arrange themselves of their own accord into the dimensions of space, time, and quality. It is the mechanism of the mind itself which imposes this order on phenomena and defines what is good and what is not good. Quality is subliminal. It is the bridge between subject and object. But the subject-object duality does not define quality, it is quality that defines subject and object. Quality is the mechanism, not the manifestation, and this mechanism is the transcendental aesthetic, which contains five dimensions: three spatial, one temporal, and one qualitative. Every phenomenon perceived by one's sense organs conforms to a definition on each of these scales in one's mind. Quality is entirely transcendental or *a priori* with phenomena conforming to it rather than it conforming to or derived from phenomena. All the incitements of desire and appetite are not derived from the world of the senses but are inherent in one's mind, *a priori*.

Some may spurn this postulate of the transcendental nature of the aesthetic and say: "God is the perfect being; all that we know of goodness is what we recognize as Godlike. He is like the sun above the clouds, pure and undefiled, and to mortal man on earth, though His light is obscured by the dark clouds of our ignorance and depravity, it is the only illumination available. We have nothing to do but strive to drive away the clouds and, in time, we will know the goodness of God through His revelations." But first let us analyze what is meant by the terms God and perfection and we will know the contradiction in this statement. One may define God as the embodiment of perfection or one may define perfection as all things that are Godlike, but one cannot define one term with the other without reasoning in a circle. One of the terms must be defined in relation to phenomena, as definitions must contain some sort of characteristics to which phenomena may conform. While a term may be defined as synonymous with another and thus take on all of its characteristics, that one must have characteristics to which phenomena can conform. If it is defined back again as the first, neither has any characteristics and thus they are not definitions. As theories are composed of nothing but definitions, theology cannot be considered proper theoretical knowledge but only mysticism unless the definition of God is given some characteristics to which phenomena may conform. Epistemology defines knowledge only in terms of theory and, if theology cannot conform itself to the definition of theory, there is no justification for giving it any standing other than false. Some might argue that it is deserving of a higher standing because of an independent theological method which is not based on epistemological observations but on more insightful observations called revelations. All that has been said above, however, may be said again concerning the terms God and revelation, for one may define God as all that is known through revelation or one may define revelations as knowledge emanating from God, but one cannot define one term with the other without reasoning in a circle. As no physical distinction between revelations and epistemological observations has ever been suggested, revelations can only be defined as knowledge emanating from God. God, therefore, cannot be defined in terms of revelations nor can He be defined as perfect if the characteristic of Godliness is contained in the definition of perfection.

Perfection is defined as "conforming absolutely to the definition of the type" or "exactly fitting the need in a certain situation or purpose." If God is defined as the perfect being, He is, out of the whole infinity of beings, the one who conforms most accurately to the definition or who fits the need most exactly. But whose definitions and needs are these except mortal man's? God's? No, for that is again arguing in a circle. For God to earn the title of perfection, He must conform Himself to man's definition and fit His actions to man's needs. Yet if a being is submissive to another's *ad hoc* definitions and purposes, it follows that He is also inferior to that being to which He submits. One must ask how a being who is inferior to another can be defined as perfect in the theistic sense of the term? If goodness were not entirely *a priori* but had an independent existence outside of our minds in the world of phenomena, how could anyone recognize what is Godlike and what is devilish? While knowledge of what is perfect requires omniscience, mortal man has only a finite set of observations to call his own. Could not a devil, if there is such a thing *a posteriori*, present himself as the best of all phenomena yet observed and then betray people as soon as he had gained their trust? Surely we would all be lost if there were not an *a priori* basis for the judgment of phenomena which applies to all phenomena and is not dependent on any finite set of observations. Where does judgment of phenomena take its certainty if all the rules which it follows are again and again induced from observation? Without certainty of judgment there would be no prevention for falling into the traps of the devil, for only with an *a priori* understanding of what is good and what is not good, abstract from any particular observations, could one foil a devil who would deceive one with false pretenses. Else one would say on the devil's first sign of largess (or even after several signs), "This thing is perfect, so it must be God" and then one would follow it and soon find oneself betrayed and left to burn in the pits of hell for one's blindness. Let us be thankful that our understanding of good and evil is of our own making and not imposed on us by any being, God or devil.

Some might respond that the father of lies must have a consistent program; that his house, if divided, could not stand. But it is apparent that in many people the desire for a leader is greater than the onerousness of suffering under the several tyrants who succeed one another promising redemption from the last, whom they claim was shortsighted in having swindled the victim so quickly, having in mind rather a longer and more thorough fleecing. Such victims wind up with nothing, not even revenge, for that was handled for them by each of the tyrants on his predecessor. If people did not have an *a priori* understanding of good and evil, they would be defenseless against tyranny, provided that the tyrant was clever enough to have more than one frontman and let them compete with each other for the favor of the people. That this does not happen (an least not for long) is evidence that it is the mechanism of the mind itself that defines what is good and what is not good.

1.2 The autonomy of the will from God

If the characteristic of omnipotence is contained in the definition of God and perfection, one must assume that the clouds referred to in the above passage are of His own making. If so, their existence depends on His will and fits His purposes or else He would not have made them. One can only ask what His purpose might be to obscure His own light? Some would respond: "It is because we do not value what is too easily given us. Only after we have striven long and hard to drive away these clouds of ignorance and depravity may we be allowed a glimpse of His light; else we would be unprepared for and unworthy of His enlightenment." Yet is not this mechanism of one's mind which values some things more and some things less and decides what shall be striven after and what not also of His making? It must be if we are to maintain the characteristic of omnipotence for, if it were not, there would be an exception to His power and He could not be considered omnipotent and thus not perfect either. If it is, His actions are clearly illogical, for why would He create a problem only so He could create a solution? If He wanted us to have enlightenment and value it too, a truly omnipotent being would simply proclaim that we be enlightened and also proclaim that we value it, thus avoiding the whole circuitous route described above. Maintaining that He does create the clouds of ignorance and depravity for us, as the characteristic of omnipotence demands, conflicts with that same characteristic. The very need to create the clouds presupposes that one's value scale (the mechanism which determines what is good and what is not good) is beyond His power and thus He is not omnipotent and therefore not perfect either. If one's value scale were within His power, He would simply alter it by fiat so that one appreciates and is worthy of enlightenment, however easily it is given, thus making the creation of the clouds of ignorance and depravity superfluous.

Some might dispel this contradiction by asserting that He may be omniscient and yet not have the power to alter one's value scale by fiat and therefore the creation of these clouds which obscure His light are the only means available to Him to assure that what we do know of goodness we appreciate. To assert that He cannot alter one's value scale by fiat (instantaneously), however, is to say that His actions, like those of mortal man's, take place only over a period of time and that He does not have the power to quicken the process but must wait patiently for events to unfold through their natural duration. Time itself, then, is the master of God to which He must conform His actions. Yet time is only a dimension in the transcendental aesthetic of man to which all phenomena must conform (along with the other dimensions) to be defined as phenomena. To say that God must conform Himself to this dimension is to say that He is a phenomenon for, by definition, all that so conform are phenomena and all that are not phenomena do not conform. If a being must conform himself to man's definitions (whether they are definitions on one's value scale as was referred to earlier or definitions on one's temporal scale as is referred to now), it follows that he is inferior to that being to which he submits. One must ask again how a being that is inferior to another can be defined as perfect in the theistic sense of the term?

One's will, which determines what shall be striven after and what not, is also contained in one's mind, *a priori*, and not in His, for the will is a function of the transcendental aesthetic. One's will is therefore autonomous, as any attempt to subvert the will results in a contradiction, for every master of the will can be shown to derive its power and its very existence from nothing else than the will itself. Every attempt at heteronomy should be seen as an attempt by an individual to place other people under his own coercive will. This coercion is maintained, not by superiority of arms which can always be resisted, but through the subtle trick of erecting an omnipotent God to which each of our wills must submit but which is really only an invention of the one individual's will. His proclamations, then, issued under the guise of omniscience, are nothing more than what has been determined as good by the value scale of the individual who erected Him.

Section 2

2.1 The Creator as a noumenon and not a phenomenon

Theology must define a God that has some material effect on our lives (a phenomenon) if it wants to command the attention of acting man. Yet the foregoing discussion has proven that any possible definition of the phenomenal existence of God leads us to contradictions and, hence, God is undefinable and has no material effect on our lives nor can we hear Him speak to us even if we listen. What ontological effect do these contradictions have on the Creator's existence? Only that It or any of Its special effects (miracles) do not appear to us in phenomena nor may we ever find It in phenomena through the invention of new theories. We cannot see behind the veil of Maya. But what is there that appears to us as phenomena without a thing appearing? This thing in itself is a noumenon, but we know nothing of it for all that we can know of is phenomena. In phenomena, empirical inference may suggest that there is an order that is not of our design (general theories), so noumena may have purposes, and such purposes can only belong to a thinking being, omniscient over all phenomena. This being is the Creator and Its omniscience is timeless, omnipresent, and amoral for time, space, and quality are products of our own minds, *a priori*. The laws that form from Its purposes are immutable and have their effects over all phenomena without regard to their position in space, time, or quality. It is vain to expect the Creator to break these laws which It has created out of Its purposes for the mere benefit of man, whether they are good men (ones who like their own actions) or not. Yet what is a miracle except the breaking of one of these immutable laws? It is ironic that miracles, so touted in the writings of churches and which are the sole support for so many people's faith, are the one thing which, if true, would falsify the belief in an omniscient God. If He were truly omniscient, would He not have foreseen the situation which calls for the breaking of the law before the formulation of the law and rather have formed a better law which would cover that situation as well as all the others? Prayers for intervention or guidance imply that God botched His creation and left it too inept to survive without help or too dull to understand without clues. The theist's God is a washed-out engineer turned tinker.

Divine laws, noumenal as they are, can never be known by man for they exist outside the dimensions of space, time, and quality within which all knowledge is arranged. The Creator's purposes do not conform to the definition of purpose or want, that is, a high position on one's value scale. The characteristic of a value scale can only be attributed to an individual's transcendental aesthetic, which organizes phenomena, and not to noumena, which creates phenomena. Nothing about the Creator, not even omniscience, for it contains the characteristic of anthropomorphic sense perception, can be defined, for definitions are only in reference to phenomena and not to noumena. To ascribe anthropomorphic characteristics such as purpose, will, or omnipotence (which is also anthropomorphic as it concerns only phenomenal power) to the Creator is to degrade Its stature to that of a mere phenomenon, albeit a very powerful one.

2.2 The conjectural nature of metaphysics

Knowledge of what exists in the mind of another cannot be directly observed. It can only be known within limits set by the actions one takes and then only on the assumption that one is exercising a certain level of competence in designing one's actions to suit one's purposes. Because of this, theories concerning human action are obscure in their application and thus one cannot take actions perfectly suited to fulfill one's purposes in any given situation. Limited as it is, this ability to know what is in the minds of other people may cause one to arrogate to oneself the ability also to know what is in the mind of the Creator. Thus, one may believe oneself to possess knowledge of noumena, which one cannot gain from epistemological theories which are only applicable to phenomena, in the same way that one tries to possess knowledge of what is in the minds of other people though it cannot be directly observed. That the Creator is referred to with anthropomorphic terms such as "mind" or "purpose" implies this comparison of It with humans. While there is no restriction on what one thinks, one cannot claim to have knowledge unless it conforms to the definition of theory, which contains the characteristic of being composed of definitions. Only phenomena can conform to definitions, so one cannot claim to possess knowledge of the Creator without contradicting oneself, for It is a noumenon. As the idea of noumena is thought of only because of people's fundamental ignorance, it entails a contradiction to speak of divine transcendence when the very idea of noumena is directly connected with man's ignorance. The division between phenomena and noumena is not just a limit among other limits one may encounter but is the absolute *limit*. One cannot be consistent and believe in a God that is known or thought of on both sides of the *limit*. The Creator is either purely noumenal or It is a living, breathing creature like any other phenomenon, though perhaps more powerful than any other.

The personification of the Creator may provide some comfort, perhaps by degrading that which is bewildering to the level of a phenomenon, just as other individuals are phenomena, and thus allowing one to best It in the same way that one can manipulate other individuals when they let their value scale become known. However, this personification must remain strictly conjectural if one is to avoid contradiction. As long as thoughts of the Creator conforming to anthropomorphic characteristics are kept purely conjectural, however, there is no harm in them and they may serve some useful purpose. The fact that faith is conjectural not only relieves it of the strictures which truth imposes on theoretical knowledge but also immunizes it from refutation. The contradiction is not in the belief that noumena may conform to anthropomorphic characteristics but in the belief that they are known and are not merely conjecture. An atheist who attempts to disprove the existence of the Creator for lack of empirical evidence is just as guilty of the above contradiction as believers who claim to have knowledge of Its existence through revelation. If the Creator is anything, It is undefinable. The only thing that a mystic or an atheist can be certain of is that they are ignorant as to the referent of their argument. Since the proof of this does not alter the effect of their belief on the faithful, however, it is irrelevant. What is important is not the truth of a religious doctrine but rather what actions are taken in its name by believers, particularly whether or not they are driven to violence. For practical purposes, God may be defined simply in terms of what actions are taken by the faithful in conjunction with their belief, so "to believe in God" is to take those actions and not necessarily to have any specific opinion concerning divine transcendence.

2.3 The definition of the term "transcendental"

The term "transcendental" employed throughout this book is synonymous with "*a priori*", which is defined as whatever exists in the mind prior to and independent of experience. The meaning here is clear: The spatial, temporal, and qualitative dimensions are not found in phenomena but are part of the mind's own mechanism for bringing order to the jumble of sense perceptions. For the reader who is not clear what exists in his mind prior to and independent of experience, this may seem like a vague definition. Only Absolute Geometry exists in one's mind prior to and independent of experience, that is, all the usual axioms of geometry with the exception of any statement regarding the uniqueness of a parallel to a given line through a particular point. Given line AB, point C not on AB, and point D such that angle ABC equals angle BCD, is line CD the same regardless of the value of this angle? Neither the affirmation nor the denial of this statement is intuitive enough to be considered transcendental. The world is Euclidean (or is a small enough region in hyperbolic space to be approximated by Euclidean Geometry), so it makes sense that this is how people organize phenomena in their minds. On the other hand, Hyperbolic Geometry is simpler because it identifies similarity and congruence. There is evidence that people use both systems at different times, being troubled by optical illusions when thinking hyperbolically. The only real difference with living in a hyperbolic space (and thinking hyperbolically) is that one can readily tell whether a phenomenon is big and far away or small and near, which is not always clear in Euclidean space.

The term "transcendental" may also refer to noumena, as it is what transcends our sense perception. The analogy is that, just as the five dimensions exist in a person's mind prior to and independent of experience, so the Creator's purposes exist in Its mind prior to and independent of experience. Are the two senses of "transcendental" identical? Is one's own mind the Creator and the material world an illusion? Perhaps. They are at least synonymous, for they both have the same extension; that is, people have to deal with the same phenomena, whatever its source. Just as surely as theists meet with insolvable paradoxes when they allow a noumenal God to conform to their definition of Him by His phenomenal effect on them, so aestheticians meet with insolvable paradoxes when they allow the criteria of quality to conform to their definition of it by its phenomenal effect on them. Both are subliminal; when one person speaks of these things and another listens, neither of them understands.

This is the mistake of aestheticians: Rather than defining a definition as being on one's value scale at a certain point, they define quality itself as something phenomena has. The value scale as a whole entity can be defined, but the criteria of why this and not that definition is high on the scale transcends understanding. Quality is not something embedded in the phenomena which conform to definitions but rather definitions are embedded in one's value scale. For the same reason, the term "definition" can be defined, but to define not the whole entity of "definition" but only specific ones is to distinguish them by their intensions, which are transcendental. To allow this and not that definition to conform to a specific definition, according to whether it has certain intensions defined by the definition in question, is to create absurdities such as Russell's paradox. The intension by which his definition distinguishes its conformant definitions is that they do not conform to themselves. This intension, like all intensions, is transcendental; it is other definitions and not phenomena which conform to Russell's definition and therein lies the root of the problem. Just as the term "quality" can be defined but what distinguishes specific instances of quality from each other cannot, so the term "definition" can be defined but what distinguishes specific instances of definition from each other cannot. The criteria of quality and the intensions of definitions are *a priori*. They cannot conform to our understanding because they *are* our understanding.

CHAPTER II

EPISTEMOLOGY

Reason, holding in one hand its principles...and in the other hand the experiment which it has devised according to those principles, must approach nature in order to be taught by it; but not in the character of a pupil who agrees to everything the master likes, but as an appointed judge who compels the witnesses to answer the questions which he himself proposes.

Immanuel Kant
Critique of Pure Reason[4]

Section 3

3.1 The three senses of the term "truth"

Whereas noumena can only be thought of, phenomena may be known, for all phenomena conform to one's *a priori* definitions. A true definition is not one which conforms to experience but rather one to which experience itself conforms. Furthermore, theories cannot be arbitrary but must be possible according to the Principle of Contradiction, independent of all experience. In another sense, however, a theory cannot be called true in reference to a certain experience if it does not apply to that experience. Hence, there are three senses of the term "truth":

1) The observation that a phenomenon exists which conforms to a certain definition, or the observation that no phenomena exist which conform to that definition;

2) The applicability of a theory to a situation so that the relations or the characteristics to which those phenomena conform can be predicted by their conformance to the other, or the contingency of that theory's application on the conformance of phenomena not yet observed; and

3) The ability of a theory to apply without contradiction to some situation, or the impossibility of that theory to apply to any situation because its every alternative contains a contradiction.

While the nominal definition of truth (the agreement of a cognition with an event) does not change, the nature of that agreement and its proof does as one crosses from an *a posteriori* sense to an *a priori* sense of the term. First, let us consider truth in its most common sense, that of empirical verification of phenomena. If one is without doubt that phenomena conform to a definition, that definition is true in the first sense. There are two criteria of first-sense truth, characteristics and relations, either of which is enough to establish a definition's truth. First-sense truths are reliable only as far as one's senses can be trusted. As one's senses are limited, no phenomenon can ever conform to an *a priori* definition with absolute certainty but always with some degree of error. Appropriate precautions should be taken to avoid being deceived by an illusion. Whenever an observation is crucial to determining the truth of a theory in the second sense, that observation should also be predicted by other theories independently of the theory in question. If a theory is considered false in the second sense because a certain phenomenon has not been observed, it is best if that phenomenon is also shown not to exist by other theories, for lack of observation carries no more certainty than observation. These precautions are especially important for observations recalled from memory or hearsay. Nevertheless, first-sense truths are never certain but always entail an error. This error may be negligible or it may be significant, depending on the situation. While it is best that crucial phenomena be predicted by several theories independently, each of these is subject to error in their application. Some theories are inherently inaccurate in their predictions, not because the theories themselves are incorrect, but because they define phenomena which are hard to verify in the first sense.

The second sense of the term "truth" is of applicability of a particular theory to observed phenomena. If all the intensions (characteristics and relations) conjoined in a definition or alternative are known from observation to be either true or false in the first sense, it is idle to determine if the theory is applicable to a given situation, for, even if it is, no new knowledge is gained. If a theory, in either all of its characteristics or all of its relations, applies to a situation, however, that situation is also known to conform to the other intensions (relations or characteristics, respectively) of the theory, thus supplying knowledge which cannot be had from observation. If the phenomena that might conform to some of the characteristics is concealed, the necessity of that situation conforming to the relations is contingent on an experiment forcing the concealed phenomena to show themselves as conformant to those characteristics. Conversely, if the phenomena that might conform to some of the relations is concealed, the necessity of that situation conforming to the characteristics is contingent on an experiment forcing the concealed phenomena to show themselves as conformant to those relations. Some of the characteristics are referenced by the relations of a theory, so phenomena's conformance to the relations implies conformance to these characteristics. The non-referent characteristics are the part of a theory in need of prediction.

The third sense of the term "truth" is applied to a theory abstract from all phenomena. Having defined, *ad libitum*, certain postulates which appear to be useful, justified only in terms of their elegance and convenience (legislative postulation), a theory is then invented entirely *a priori*, though in such a way that it is possible according to the Principle of Contradiction. To test the Principle of Contradiction, a theory is progressively transformed so that each step is equivalent to the others (they all have the same extension or are true in exactly the same situations and no others) and the last step is in the form of alternational normal schemata (truth tables). This form partitions into one or more disjunctive groups the conjunctively-joined definitions representing each situation to which that theory might apply. If, in any alternative (clause), an intension is at once affirmed and denied, that alternative is contradictory and may be dispensed with, as the Principle of Contradiction states that phenomena never both conform to an intension and not conform to that intension. A theory is possible (true in the third sense) if not all of its alternatives are contradictory. If they are all contradictory, it is nonsensical to concern oneself with whether that theory applies to a certain situation, as it has already been proven to apply to no possible situation.

If a theory contains *n* definitions with each one having phenomena either conform to it or not, there are 2^n possible situations to which one may consider applying the theory. Phenomena can at once conform to one and only one of these 2^n alternatives; phenomena cannot conform to more than one without at once conforming to the affirmation and the denial of a definition. If a theory has been invented in such a way that one or more of these 2^n possible situations can conform, even if they have never been observed, the theory is true in the third or abstract sense. It cannot be refuted in this sense by contradictory phenomena for the contradiction would be caused by factors not defined by the theory. This would only prove that the theory in question is inappropriate for the particular phenomena and is thus untrue in the second sense but not in the third. If a characteristic is implicit in a definition, the ontological question, "Is there a phenomenon conformant to that definition which does not have this characteristic?" is idle for, by definition, if a phenomenon does not have that characteristic, it does not conform to the definition. A theory may be proven untrue in the third sense only if, after all traces of empirical phenomena are abstracted from the *a priori* definitions, a contradiction remains in the definitions themselves. If this is the case, new definitions must be sought with at least one alternative that contains no contradiction. While such definitions should have applicability to observed phenomena to be useful in practice, it is not a necessary criteria. A theory, however clearly and distinctly it is conceived, may be true in the third sense and yet be academic and have no applicability to observed phenomena and thus have no observed instances of truth in the second sense, but only possible ones. Phenomena cannot be forced into the procrustean bed of a theory merely to give it applicability.

3.2 The intension of a definition or alternative

Every definition contains intensions (characteristics or relations) to which phenomena must conform to make that definition true in the first sense. Characteristics are anything that can be used to distinguish one phenomenon from another. Every characteristic conjoined in a definition is self-contained and does not refer to any other characteristics outside itself. If phenomena (whatever their position, size, or quantity) conform to all of the characteristics in a definition, that definition is true in the first sense. Other intensions of a definition do not define characteristics, however, but refer to the relations of characteristics with other independently defined characteristics. These are called the relations of a definition and, because they refer to something not contained in the characteristics (other characteristics), they independently verify first-sense truth. Relations may take three forms, each of which have a different sense when a definition is thought of on the spatial, temporal, or qualitative dimensions. They are as follows:

relation	space	time	quality
position	distance	date	importance
size	length	duration	vagueness
quantity	mass	frequency	potency

Position is the relation of a definition situated at a particular place on a dimension with the position of another that is arbitrarily defined as the origin. There is no inherent reason why one definition or another should be thought of as the origin, for the characteristic of origination is not contained in the definition of the transcendental aesthetic dimensions. This choice is made with regard to convenience only. To be useful in practice, an origin must be such that the phenomenon that conforms to it is very stable, never conforming to definitions at other positions on that dimension. While distance and date can contain the intension of magnitude, the importance of a phenomenon can be known only as greater than or less than another. For importance, exact spatiotemporal magnitudes are replaced with vague terms such as "a lot more" or "a little more."

Size is the number of units along a dimension from where a definition first appears and where it leaves off, the unit being the same as the one that measures position. As such, there are phenomena that can be used to measure length and duration but not vagueness. As the name implies, the vagueness of a definition is determined by one's inability to determine the exact multiple of a unit (of money, for instance) which is equivalent to the definition in question. Such knowledge is essential if one proposes to exchange a phenomenon that conforms to that definition with phenomena that conform to a number of those units so that, relative to one's own value scale, what one acquires conforms to a higher definition than what one surrenders. Relative to one's trading partner's value scale, the other has more value. That two people might have such opposite valuations is not surprising when one considers the instability that diminishing utility gives to all phenomena.

Quantity is not measured in the same units as position and size, though potency is unstable for the same reason that importance and vagueness are. The unit of quantity is the stime or the smallest bit of phenomena which still conforms to a definition. Multiples of this unit determine the mass, frequency, and potency of phenomena. Potency, like distance, duration, and the others, is transcendental and can thus be defined at infinity if one wishes, for the phenomenal world is no constraint at all on theoretical knowledge. Defining potency at infinity, a step that is needed to dispute certain arguments for the existence of God, is equivalent to affirming the existence, but not necessarily the uniqueness, of parallel lines. The existence of parallels depends on the exact definition of the term "transcendental." See section 2.3.

As with position and size, a definition need not imply quantity; that is, there may be no other characteristic whose relation to the given one is stated. Where there is such a characteristic, it must be contained in each definition or alternative of a theory so that the theory is self-supporting. Relations are usually only between the characteristics of a theory, though sometimes a characteristic common to many theories such as origin, unit, or stime is used for measurement.

Section 4

4.1 Linguistics and rhetoric in theoretical writing

If a theory is to be used by more people than the author, its definitions must be expressed in words common to all literate people. The vocabulary of a society is nothing more than the definitions of phenomena observed and predicted by its theories, and grammar is the means to make clear the logical connections between the definitions. It is important in all theoretical writing to import no unexplained meaning to words that they do not ordinarily contain and to use grammar that is consistent with the precepts of formal logic. What distinguishes formal logic from the vocabulary and grammar of rhetoric is the complexity of individual terms (both definitions and logical relations) relative to the complexity of statements containing those terms. Discourse in a language is made easier by using terms that contain much meaning, for it allows shorter statements to convey a particular thought. Discourse about a language is made easier by using very simple terms, though they produce statements that are too lengthy and complex to be used for communication. This duality of purpose is solved by the simultaneous use of two languages. Formal logic fulfills the purpose of facilitating analysis but fails as a device for casual communication. Rhetoric has the perspicuity to fill this need while proving ambiguous with regard to analysis. Each is translated into the other whenever the purpose at hand requires it, so neither can be considered the foundation to which the other is reducible or from which it derives its true meaning.

The problem of translating between rhetoric and logic is the subject of linguistics, which must find ways for people to express their *a priori* definitions and theories in words whose meaning can only be communicated through ostension. One does not think in words. One's *a priori* definitions are just visions and images in one's mind's eye that do not necessarily have any words or phrases associated with them. Understanding is not a function of one's ability to explain what one knows. Explanation is just something people do who are inclined to teach. Whoever is accomplished in an art or science probably consults a lexicographer when naming his definitions so that he may be assured that as many people as possible sympathize with his understanding. If he has no concern for communication, he may dispense with the whole of rhetoric, his theory remaining true or false, as it were, though people have to judge it by its effects and not by its explanation. Simply consulting a lexicographer, however, does not assure that people will understand one's definitions, for a lexicographer can only point out what phenomena conform to those definitions and only rarely do phenomena conform with such accuracy as to conjure up the image of each intension of a definition and with such austerity that no other intensions are suggested.

Clear writing should not be construed to mean simple writing. Some readers have no talent for abstract thinking and blame all of their difficulties on the writing style of scientists, comparing it to that of bureaucrats, which is admittedly opaque. It goes without saying that to write down to such people can produce nothing but mediocrity, both in thought and in exposition.

Many people are not very adept at inventing theories to which their environment conforms and which make useful predictions about that environment and the effect of actions taken in it. They are, however, skilled in manipulating words and have endeavored to elevate linguistics above understanding. "Whatever cannot be explained in our language is not true" proclaims dialecticians, and then they quiz an accomplished individual for an explanation of his science. This is a trap, for it is a game that cannot be won while one can easily become ensnared in one's own words, at which point the dialectician will say, "I was only asking questions - if this sophist could not answer them, he must be an impostor." Conversation, especially with intellectuals, tends to make small of its subject and is something to be avoided.

4.2 The definition of the term "abstract"

Conceptus abstractus, the most common use of the word, denotes, as a verb, the act of removing from a series of perceptions that which is similar or, as a noun, the characteristic thus removed. This abstract intension is considered the essence of the phenomena; a part of the whole is substituted for the whole, the whole being concrete. Even though no pure phenomena may ever be found, and in fact may be known not to exist, the intension (characteristic or relation) remains empirical as it is derived *a posteriori*. One may believe oneself to possess knowledge of what is essential, though not necessarily knowledge of what actually is, that remaining to be described as accurately as possible from direct observation.

If similarity is thought of as residing in phenomena so that one may know the essence of phenomena by describing what is similar in them and disregarding all that is dissimilar, one must assume that the Principle of Similarity entails the distinguishing mark of what is meaningful. This assumption is based on a confusion between the techniques of rhetoric and epistemology. Because so much of what people know is learned from the rhetoric of others who have already proven their choice of definitions to be meaningful and their theories to be sound, students associate rhetoric with original thinking as well. In teaching an established theory, one who is accomplished in it gives many examples and analogies so that students can learn the meaning of his definitions through induction, by noting what is similar or indiscernible, and then forming their own definitions by identification of the indiscernibles. This method of learning, however, when applied to nature, proves false, for there is no guarantee that, because some phenomena are similar in one respect and also conform to a certain definition, other similar phenomena also conform to that definition. Even with an accomplished individual demonstrating a well-established theory with examples of what phenomena conform to its definitions and analogies of how they are related, it is a difficult task for a student to make the theory his own. How much more difficult, even impossible, the task when there is no such instructor, for nature is no *deus ex machina* that sends us prepackaged observations that we need only take note of and watch converge into a coherent theory. Such an explanation attempts to explain knowledge, which is only problematical, with God, Who is conjectural.

One cannot idly observe and wait for indiscernibles to identify themselves as meaningful definitions; one must force nature to reveal itself by performing experiments that leave it no option but either to conform or not conform to a particular theory that one has established beforehand. The identification of indiscernibles can never produce original theory for, as each intension is dropped from a description, presumably making it describe a more essential nature of phenomena by broadening its application, more things are disregarded for being dissimilar. Thus, the most general description, which should describe the most essential nature of phenomena, in fact disregards all intensions of those phenomena. Such a general description is meaningless, for to have meaning one must be able to deduce specific cases from the general rule. One cannot do that here, however, for to arrive at such a general description all specifics have been successively forgotten. Reliance on the Principle of Similarity to reveal the common and crucial elements of phenomena can produce no other result, for it is entirely dependent on a certain forgetfulness that allows one to disregard some intensions of phenomena but not others. If phenomena are remembered, not in mere outlines but in their entirety, they cannot be considered exactly similar without *a priori* knowledge of what can be considered nonessential and thus forgotten. To reason by analogy is to note that several phenomena conform to a certain definition and to forget their other characteristics and, then, when those other characteristics become important, to assume that they are also similar because the phenomena are similar in their remembered characteristic. People who equate understanding with their ability to analogize every event with one from their past tend to be less than wise. And, if understanding cannot grow out of similarity, the verisimilitude of storybooks is an even more barren ground.

What is really meant by the Principle of Similarity is not that intensions are found *a posteriori* in phenomena by noting what is similar, but rather that a theory is tacitly assumed to be meaningful (as one can usually do with established theories) and then phenomena are observed to see if any of them conform to the theory. All that do conform to it are deemed similar, giving the theory second-sense truth in those cases. Such similarity only has meaning because the *a priori* definitions do, and their meaningfulness is derived, not from the Principle of Similarity, but because they were invented to be meaningful.

The advantage of age and experience is that past events may have prompted the invention of correct theories that can be used again and again throughout life. Newcomers may not yet have had occasion to invent the needed theories for their program and must stop and think their way through problems before proceeding, where more experienced people have memorized the solution. But simply analogizing current events with past ones is meaningless; nothing prevents young fools from growing into old fools.

Conceptus abstrahens, the use of the term "abstract" employed in this book, is not something induced out of phenomena but is rather something pure, in the abstract, untainted with empirical observation. A theory is abstract for, while it may apply to many different situations, it does not derive its possibility from any of their existences. It is not disturbed by any other characteristics to which the several situations might conform and by which they distinguish themselves from one another, as long as those characteristics are not defined as mutually exclusive with that theory's definitions. Phenomena are similar only by virtue of their general conformance to what is already known in the abstract. This abstract knowledge cannot emerge from a cauldron of empirical observations but must be invented *a priori.* If the theories that one uses derive their truth from nothing but previous experience, how could one speak with certainty concerning experience without falling into a *regressus in infinitum* in which each statement is derived from previous statements of a similar kind? If a theory is to be used in practice to support the truth of statements in the second sense concerning phenomena whose relations are concealed from the senses and cannot be directly verified in the first sense, it must have certainty in all situations where phenomena conform to the characteristics (or *vice versa*, where one tests the relations in order to predict the characteristics). This certainty, that where one (characteristics or relations) proves true the other is known, must be entirely abstract and untainted with empirical observation. If it were derived from a finite set of empirical data, namely what has already been observed, how could it have certainty concerning concealed phenomena, which is where it is needed the most, as such phenomena cannot be directly verified in the first sense? Only when a theory is invented *a priori* in such a way that it is confirmed by the Principle of Contradiction can it have certainty over all phenomena that conform to it in the prescribed manner. Truth in the second sense is often confused with that in the third sense, leading some to believe that a theory's possibility is not entirely *a priori* but rather relies, at least in part, on empirical data. The operation of determining whether a theory is true in the third sense by finding at least one of its alternatives that contains no contradiction must be kept strictly separated from determining whether it is applicable to phenomena and thus true in the second sense. For people immersed in practical affairs, an academic theory that is true in the third sense, yet has no observed instances of truth in the second sense and thus no applicability, is as good as false. This leads to the belief that a theory is not possible until it is tested against phenomena. While empirical observation is often used during the invention of a theory to help guide its construction into areas of immediate practical use in cases where an empirical problem arises before a theory is fully developed, this does not mean that its invention is dependent on those observations.

Observations of whether or not phenomena conform to the definitions of a theory are normally taken as they are needed during the process of determining the applicability of the theory. It may happen, however, that an observation has already occurred (for whatever reason) and need only be recalled from one's memory or hearsay; indeed, some observations can only have been taken in the past or by other people and cannot be repeated. The fact that observations occurred before their need arose can in no way be construed to suggest that a theory was induced from them or from "experience" in general. The development of theories is never a finite process; more and more subtle implications can always be discovered. As this requires effort, it is often new empirical problems which prompt such invention. This should not detract from the *a priori* nature of what is learned, however.

The confusion between the different senses of truth is partly due to the ambiguity of certain words. The term "truth" is one such word. Another is "existence". A common use of the word “existence” refers to first-sense truth, that is, the observation that a phenomenon exists which conforms to a certain definition. Indeed, the definition of first-sense truth (see section 3.1) includes the word "exist." However, if a book is concerned only with phenomena that conform to a certain definition *p*, the author may write "*q* exists" to affirm the third-sense truth of *p* and *q*, that is, the ability of *p* and *q* to apply without contradiction to some situation. Whether or not this ability is realized and the theory has second-sense truth is a strictly separate question. This use of the term "existence" is particularly common in mathematical books where p defines a number system. In general, mathematical books, including this one, are about the real numbers unless they explicitly state otherwise. Thus, when I say "Now, the existence of a saturation point for every phenomena will be proven" I intend to prove only that real numbers and saturation points are not inconsistent. That my theory of economics has second-sense truth and the term "saturation point" means anything at all depends on whether or not the reader accepts the theory's three axioms. For that matter, the reader may not accept the existence of real numbers. Real numbers include irrational numbers, whose very name recalls an age when people did not believe that the phenomenon called "number" conformed to such a definition. There will certainly be people who do not believe that the phenomenon called "value" conforms to the definitions of total ordering, marginal utility, and proportionate effect, which is all that my three axioms assert. Either way, it still makes sense to prove that saturation points exist and other theorems as well. Real analysis and economics are both variations of a game called "Prove it!" and this game can be enjoyably played by everyone, whether they accept the existence of the phenomena in question or not.

4.3 A three-term system of formal logic

A new system of formal logic will now be introduced. The three terms of this system of logic are `P` for possible, `I` for impossible, and `M` for maybe (similar to Zen Buddhism's *mu*). Following are eleven logical relations concerning the definitions *p* and *q*. These statements are followed by a truth table which shows, in each of the four situations with which one could be presented when observing phenomena's conformance to *p* and *q*, whether the statement affirms its possibility, its impossibility, or says nothing about that situation. Except for the last one which is usually not reversed (its reversal is reversal), they are listed in pairs so that what an odd-numbered statement affirms and denies, the following even-numbered statement denies and affirms, respectively. This list does not exhaust all possible logical relations, only the most common ones.

Do phenomena conform to alternative *p*?	T	T	F	F
Do phenomena conform to alternative *q*?	T	F	T	F
1) *p* is possible	P	P	M	M
2) p is impossible	I	I	M	M
3) the possibility of *p* implies the possibility of *q* all phenomena that conform to *p* conform to *q* if *p* is possible then *q* is possible *p* is possible only if *q* is possible *p* is not possible unless *q* is possible *p* is not possible except where *q* is possible *p* is a sufficient condition that *q* is possible *q* is a necessary condition that *p* is possible	P	I	M	M
4) the possibility of *p* implies the impossibility of *q* all phenomena that conform to *p* do not conform to *q* if *p* is possible then *q* is not possible *p* is possible only if *q* is not possible *p* is possible unless *q* is possible *p* is possible except where *q* is possible	I	P	M	M
5) *p* and *q* are compatible. p and q are possible together it is possible that a phenomenon could conform to *p* and also to *q*	P	M	M	M

6)	it is impossible that a phenomenon could conform to *p* and also to *q* no phenomenon can conform to *p* and *q* together *p* and *q* are mutually exclusive	I M M M
7)	*p* and *q* are both possible either *p* or *q* is possible	P P P M
8)	*p* and *q* are both impossible neither *p* nor *q* are possible	I I I M
9)	*p* and *q* are only possible together *p* (*q*) is possible if and only if *q* (*p*) is possible *p* (*q*) is a necessary and sufficient condition of *q* (*p*) all phenomena that conform to *p* conform to *q* and all that conform to *q* conform to *p*	P I I M
10)	*p* and *q* are both possible, but not together either *p* or *q* is possible, but not together	I P P M
11)	*p* and *q* are equivalent *p* and *q* are synonymous	P I I P

By marking a P or an I only where one positively states that an alternative is possible or impossible, one never has to make assumptions about what is meant. Where mainstream logicians have to guess at the meaning of "if *p* is possible then *q* is possible" in cases where it is known that *p* is impossible (they guess P), this system leaves those alternatives at M. This is as it should be, since no statement has been made concerning their possibility. Furthermore, "unless", which mainstream logic translates as synonymous with "or", is translated more naturally by the new system. Other ambiguities are also cleared up, but more importantly, the great crack in the foundation of mainstream logic where first-sense and third-sense truth are confused has been resolved. Whenever mainstream logic speaks of affirmation they refer to phenomena having been observed that conform to a definition (truth in the first sense) and whenever they speak of negation they refer to the impossibility of phenomena conforming to a definition (falsity in the third sense). The three senses of truth must be strictly separated; there is no direct connection between what people assert and what exists in reality. People often lie and, when they do, they try to be consistent. Figure 4, taken from **Methods of Logic**[5] by Dr. Willard Quine, illustrates the problem. Here, whiteness signifies lack of information (maybe), shading signifies emptiness (impossibility), and a cross signifies nonemptiness (observation of conformant phenomena).

Figure 4

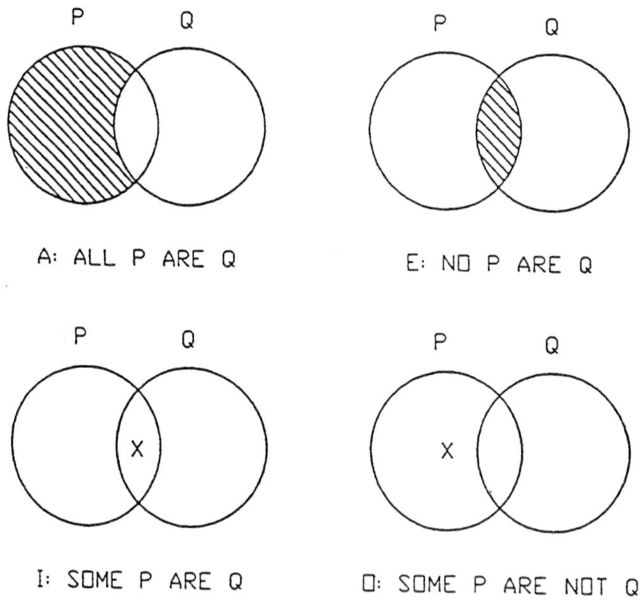

Now let us consider the case (figure 5) where p is entirely impossible. Comparing this diagram to figure 4, it is apparent that the statement "some p are q" is false while the statement "all p are q" is true. This is an absurdity; if one were to take the witness stand in a court of law and state that all p are q while knowing full well that there are no p, q or otherwise, one would be committing perjury. Most people would take one's statement to imply that p is possible. A more mathematical example is the statement "If x is a real number such that $x^2 + 1 = 0$, then $x = 0$." Most people take this to mean that zero is the only real number such that $x^2 + 1 = 0$. This is not true because zero squared plus one is one, not zero. In fact, there is no real number such that $x^2 + 1 = 0$. However, in mainstream logic any conclusion can be attached to a false premise and the implication is still considered to be true. In general, let the conclusion be $x = c$ with c any (fixed) real number.

Figure 5

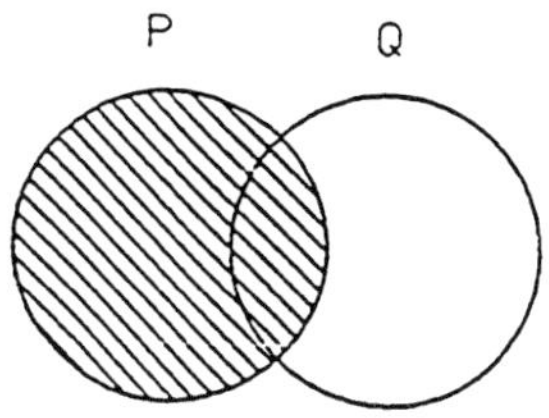

THERE ARE NO P

$x^2 + 1 = 0$	Given.
$(x - c)(x^2 + 1) = 0$	Multiply both sides by $(x - c)$.
$x - c = 0$	Divide both sides by $(x^2 + 1)$. Because x is real, $x^2 + 1 \neq 0$.
$x = c$	Add c to both sides.

This is absurd: We are told that x is real and prove that x is any real number c. In other words, the relation $x^2 + 1 = 0$ imposes no restrictions on x whatsoever. But, in fact, this relation excludes all real numbers. Now let us consider a relation that is only slightly less restrictive,

$x^2 = 0$, which excludes all real numbers except zero. The proof given above does not work because division by x^2 is not allowed. However, taking the square root of both sides proves that zero is the only real number such that $x^2 = 0$, which is true and is what most people take that statement to mean. Thus, in mainstream logic, as a relation shifts from virtually exclusive to completely exclusive, the result jumps from one true answer to an infinity of false ones.

The new system introduced here resolves this problem by making the statement "all *p* are *q*" synonymous with "if *p* then *q*" (see #3 above), a route that mainstream logic could not take, as it separates truth-functional statements (of which "if *p* then *q*" is one) from quantificational statements (of which "all *p* are *q*" is one). To illustrate the ease of using a system where truth-functional and quantificational statements are resolved with the same method, I will use the crowning example of Dr. Quine's own treatise, **Methods of Logic**, pg. 196:

Premises: **1)** The guard searched all who entered the building except those who were accompanied by members of the firm.

2) Some of Fiorecchio's men entered the building unaccompanied by anyone else [unaccompanied by non-Fiorecchio men].

3) The guard searched none of Fiorecchio's men.

The following definitions will be used, all of which tacitly contain the characteristic of concerning only people who enter the building:

f) People who work for Fiorecchio.
m) Members of the firm.
s) Members of parties that were searched.
u) Members of parties of only *f* (Fiorecchio men).

With four definitions there are 2^4 or sixteen possible alternatives, each of which is represented by a column in the following truth table:

	1	2	3	4	5	6	7	8	9	10	11	12	13	14	15	16
Is *f* realized?	T	T	T	T	T	T	T	T	F	F	F	F	F	F	F	F
Is *m* realized?	T	T	T	T	F	F	F	F	T	T	T	T	F	F	F	F
Is *s* realized?	T	T	F	F	T	T	F	F	T	T	F	F	T	T	F	F
Is *u* realized?	T	F	T	F	T	F	T	F	T	F	T	F	T	F	T	F

While in this example each row questions phenomena's conformance to only one definition, it is sometimes useful to conjoin definitions on a row so that a "T" signifies that phenomena conform to all the definitions and an "F" signifies that phenomena fail to conform to some of them. Which definition phenomena fail to conform to is not implied by the negation of that conjunction, however. It is incorrect to assume that denying a conjunction is to say that the verb fails, thus indirectly implying that the noun does not.

First, note that, while definitions *f*, *m*, and *s* contain only characteristics, definition *u* contains a relation with *f*, so immediately we can fill in an I at alternatives 9, 11, 13, and 15 for we know that, if one does not work for Fiorecchio, one cannot be in a party of only Fiorecchio men. The remaining alternatives are still M, for nothing has yet been said about them. The first premise conforms to the #4 statement in the list of logical relations given above with *s* in place of *p* and *m* in place of *q*. Therefore, we fill in an I at alternatives 1, 2, 9, and 10, a P at alternatives 5, 6, 13, and 14, and an M everywhere else as the first premise does not mention whether any of the other alternatives are possible or impossible. The second premise conforms to the #5 logical relation, so we fill in a P at alternatives 1, 3, 5, and 7, the rest being M. The third premise conforms to statement #6 and hence we fill in an I at alternatives 1, 2, 5, and 6, with the remainder unmentioned and thus left at M. The truth table now looks like this:

	1	2	3	4	5	6	7	8	9	10	11	12	13	14	15	16
Is *f* realized?	T	T	T	T	T	T	T	T	F	F	F	F	F	F	F	F
Is *m* realized?	T	T	T	T	F	F	F	F	T	T	T	T	F	F	F	F
Is *s* realized?	T	T	F	F	T	T	F	F	T	T	F	F	T	T	F	F
Is *u* realized?	T	F	T	F	T	F	T	F	T	F	T	F	T	F	T	F
relations	M	M	M	M	M	M	M	M	I	M	I	M	I	M	I	M
1st premise	I	I	M	M	P	P	M	M	I	I	M	M	P	P	M	M
2nd premise	P	M	P	M	P	M	P	M	M	M	M	M	M	M	M	M
3rd premise	I	I	M	M	I	I	M	M	M	M	M	M	M	M	M	M

To get the result, recall that M signifies a lack of knowledge, so any statement about the possibility or impossibility of an alternative is stronger than a "maybe." Furthermore, hearing from one source that an alternative is possible cannot redeem it if one knows it to be impossible for another reason, so I is stronger than P if they both appear in the same column. Now, filling in an I wherever we see one, a P wherever we see one that is not dominated by an I, and an M, only where no statement is made either way, we get the following result:

	1	2	3	4	5	6	7	8	9	10	11	12	13	14	15	16
Is *f* realized?	T	T	T	T	T	T	T	T	F	F	F	F	F	F	F	F
Is *m* realized?	T	T	T	T	F	F	F	F	T	T	T	T	F	F	F	F
Is *s* realized?	T	T	F	F	T	T	F	F	T	T	F	F	T	T	F	F
Is *u* realized?	T	F	T	F	T	F	T	F	T	F	T	F	T	F	T	F
relations	M	M	M	M	M	M	M	M	I	M	I	M	I	M	I	M
1st premise	I	I	M	M	P	P	M	M	I	I	M	M	P	P	M	M
2nd premise	P	M	P	M	P	M	P	M	M	M	M	M	M	M	M	M
3rd premise	I	I	M	M	I	I	M	M	M	M	M	M	M	M	M	M
result	I	I	P	M	I	I	P	M	I	I	I	M	I	P	I	M

From the result, one can test the truth of any conclusion one is interested in. In this case, we want to know if the statement "Some of Fiorecchio's men are members of the firm" is implied by the premises. As it conforms to the #5 logical relation given in the list above, we must find a P among alternatives 1, 2, 3, or 4 to prove it. As the third alternative is possible, the conclusion is proven. If we were interested in knowing whether the statement "All of Fiorecchio's men entered the building unaccompanied by non-Fiorecchio men" is implied by the premises, we would need a P among alternatives 1, 3, 5, and 7 and all of the alternatives 2, 4, 6, and 8 to be impossible, as this statement conforms to the #3 logical relation. We find that alternatives 3 and 7 are possible, fulfilling the first part of the requirement, but alternatives 4 and 8 are still unknown, so the conclusion is not proven; it is a maybe. This is a more insightful "maybe" than we had before analysis, however, as we now know where our investigation must lead. If further information can be found that shows alternatives 4 and 8 to be impossible without also showing alternatives 3 and 7 to be impossible, this conclusion may yet be proven. It is contingent on an experiment forcing the concealed phenomena to show themselves as conformant to these characteristics. If alternatives 1, 3, 5, and 7 were all impossible from the information given, no further information could help us, as nothing redeems an impossibility.

For contrast to this method of logic, I will now give the method of mainstream logic as it is presented in Dr. Quine's treatise (pg. 199). Briefly, the premises are the three given and the negation of the conclusion being tested. $\forall x$ is read "for all x", $\exists x$ is read "there exists x such that", and $\rightarrow$ is read "implies." - is negation. Thus, the first premise is read, "For all *x* there exists *y* such that F*x* but not G*x* implies H*xy* and J*y*." The negation of the conclusion, "There exists *x* such that both K*x* and J*x*", is "For all *x*, it is not true that both K*x* and J*x*". The dots help clarify groupings. Dr. Quine writes (pg 29-30), "They may be thought of as a systematic counterpart of the practice in ordinary language of inserting 'else', 'also', etc.…Any group of dots alongside any of these connectives [$\rightarrow$, $\leftrightarrow$, etc.] represents a greater break than is represented by the same number of dots standing alone as a sign of conjunction, but a lesser break than is represented by any larger group of dots." The several instances combine these premises until one arrives at a contradiction and thus proves the conclusion in question or, more precisely, disproves its negation. Try it.

Fx) x is a person who entered the building
Gx) x was searched by the guard
Hxy) x was accompanied by y
Jx) x was a member of the firm
Kx) x was one of Fiorecchio's men

Premises: $\forall x \exists y(Fx \cdot - Gx \cdot\!\rightarrow\!\cdot Hxy \cdot Jy)$
$\exists x \forall y(Kx \cdot Fx \cdot Hxy \rightarrow Ky)$
$\forall x(Kx \rightarrow - Gx)$
$\forall x -(Kx \cdot Jx)$
Instances: $\forall y(Kz \cdot Fz \cdot Hzy \rightarrow Ky)$
$\exists y(Fz \cdot - Gz \cdot\!\rightarrow\!\cdot Hzy \cdot Jy)$
$Fz \cdot - Gz \cdot\!\rightarrow\!\cdot Hzw \cdot Jw$
$Kz \cdot Fz \cdot Hzw \rightarrow Kw$
$Kz \rightarrow - Gz$
$-(Kw \cdot Jw)$

Notice that the method of mainstream logic only proves one conclusion at a time while, with the result of my method, one can test the truth of any conclusion one is interested in. In mainstream logic, if one is interested in drawing further conclusions from the premises given, one must repeat the entire process. Furthermore, in mainstream logic, if one fails to prove one's particular conclusion, one neither knows where one's investigation must lead, nor can one be certain that it is impossible, for one may only have been blind to the proof. The method of mainstream logic is not algorithmic and one is actually blind to this type of proof quite often as it is more than a little opaque.

Most people have an intuitive understanding of what constitutes a proof. When a manual about methods of logic defines a new concept on virtually every page, one feels that the logician's problem is not just one of opacity. It is well known among FORTH programmers that, when one cannot think of insightful names for all the variables one must define for an application, the compiler should have provided a better temporary storage system for intermediate results. The same is true in mathematics or logic: One should never be required to attach meaning to more symbols than is absolutely necessary. For example, in first-semester calculus the derivative is sometimes defined as $\lim_{x\to a}\frac{f(x)-f(a)}{x-a}$. This definition only yields the slope at a specific point, a. Then, when one wants to derive the equation for the slope at an arbitrary point, x, one is given the formula, $\lim_{t\to x}\frac{f(t)-f(x)}{t-x}$. Three variables are named and x, which is the only one that appears in the result, changes its meaning from one equation to the next. It is better to define the derivative as $\lim_{h\to 0}\frac{f(x+h)-f(x)}{h}$ with h a "very small" number. Only two variables are named, neither of which change their meaning. The limit is easier to evaluate and h is eliminated in the course of evaluation, just as one expects of a very small number. One is left with the equation for the slope at an arbitrary point, x, which is the purpose of the exercise, not the slope at a specific point, good only for one problem. Also, the picture is easier to describe as one does not have to perform a subtraction to get the "run" part of "rise over run" for slope, it is just h. Exceptional cases where this does not work can be relegated to the exercises.

Too many variables confuse an issue. This is particularly a problem among economists, who often do not give their variables single letter names. Without an introduction to the notation, most people who pick up an economics text think that MPC (actually C'(Y), the derivative of consumption with respect to national income) means M (money) times P (prices) times C (consumption), all of which appear elsewhere in the same equation. Spelling words with variable names is one of the surest signs of someone without mathematical maturity. Also, who can remember which way the subscripts go in $MRTS_{AB}$? Mainstream microeconomics defines this and several other terms, the equality of any two being sufficient to prove profit maximization. But, if any two are sufficient, why not take two, say $\frac{dx_a}{dx_b}=\frac{p_a}{p_b}$ and omit the others? Another concept even closer to the center of mainstream economics is that of perfect competition. It is usually given a dozen or so "characteristics", but closer examination reveals that it only has four, the others being consequences of those four.

Section 5

5.1 Methods of analytic deduction and proof

Analytic deductions are all that are contained in the definitions of a theory, though they are stated to emphasize one intension or another. Analysis is the method used to form a new theory that is implied by an existing theory, but perhaps with greater extension. A syllogism is analytic because it only restates the intensions in another way than they were originally stated. There are usually several ways that an instructor can originally state his intensions and these are called discursive postulates. For the benefit of his students, the implications of the theory are then deduced from the discursive postulates. These several deductive systems are called interpretive as they can be translated into one another. In reality, there is only one system and it is not originally stated in any way, for it is timeless. Discursive postulation is a rhetorical technique and interpretive theories are identical. Any declarative sentence may be taken by linguists as synthetic or analytic as it is related to discursive postulates from which it can be deduced.

The statement "p implies q" is synonymous with the statement "if p then q", the only difference being in rhetorical usage. When speaking of theories implying other theories, the one in the "p" position is called the premise and the one implied by it in the "q" position is called the conclusion. Wherever the premise is possible, the conclusion is also, though it can be possible in other situations too. Because the only time that p does not imply q is when p is possible and q is not, the conjunction of p with the negation of q shows in each of its alternatives the several instances where q is impossible while p is possible, which thus prevents p from implying q. If p-q is inconsistent (each of its alternatives contains a contradiction such as r-r), there is no situation where q is impossible while p is possible and so the implication of q by p is proven by the contradiction method. This is useful whenever q is a more nebulous concept than its negation (Infinity, large or small, is like this). Another method of proving that p implies q is the contrapositive method. Because, whenever *q* is impossible, *p* is also impossible (*p-q* falsifies the statement "*p* implies *q*") but *p* may be impossible at other times too, the denial of *q* implies the denial of *p*; that is, the statements "*p* implies *q*" and "-*q* implies -*p*" are synonymous. If it is more convenient to deal with "-*q* implies -*p*" rather than "*p* implies *q*", its proof also proves the implication of *q* by *p*. One lists all of the instances where the conclusion is impossible and then sees if the premise is also impossible in those situations.

"*p* implies *q*" is not impossible whenever *p* is not observed or *q* is observed, so to prove that $-p \vee q$ ($\vee$ means "vel" or "inclusive or") is tautological (it is possible in all 2^n situations) is to prove that *p* implies *q*. This proof may take two forms: A check of whether $-p \vee q$ is possible in each of the 2^n situations to which it could be applied, or an analysis of $-p \vee q$ showing that it contains the disjunction of the affirmation and the denial of a definition (such as $r \vee -r$).

From the definition of implication, the conclusion, *q*, is possible in each instance where the premise, *p*, is possible, so another method of proving implication is to list all of the instances where the premise is possible and then see if the conclusion is also possible in those situations. If it is, the implication is proven. When the premise is possible in one and only one instance and it can be shown that the conclusion is also possible in that instance, the proof is called a fell swoop. This term also applies to the contrapositive method when the conclusion is impossible in one and only one instance and it can be shown that the premise is also impossible in that instance. This is rhetorically opposed to the full sweep, that is, testing whether the disjunction of the denial of the premise with the conclusion is tautological. This makes four proofs of implication. For *p* implies *q*, they are:

1) *p-q* is inconsistent;
2) *p* is impossible whenever *q* is impossible;
3) $-p \vee q$ is tautological; and
4) *q* is possible whenever *p* is possible.

A fifth proof is deduction. Deduction is the most straightforward but often the most involved method of proving implication. When a theory is in the form of alternational normal schemata, an implication is formed by either removing (deducting) superfluous definitions from an alternative to give it more extension (the fewer definitions that phenomena have to conform to in an alternative, the more phenomena conform) or by adding new alternatives. If an alternative contains n definitions, there are a finite number (2^n-1) of theories that can be implied by the deduction of definitions from that alternative. n, however, may be increased to enumerate not just the definitions but all of the intensions in an alternative. Each definition contains a finite number of intensions which are conjoined so that the truth of a definition depends on the truth of each of its characteristics or each of its relations and the truth of an alternative depends on the truth of each of its definitions. Thus, the truth of an alternative depends on the truth of each of its characteristics or each of its relations. How they are grouped in definitions is immaterial; *pq* might imply *r* if *r* is composed of some of the intensions of *pq*.

5.2 The possibility of synthetic *a priori* knowledge

While there are a finite number of theories implied by a given one, there is an infinite number of theories that could have implied a given theory. These theories are found by adding definitions (which are always available) to an alternative of the given theory. If, out of this infinity of theories, the analysis of one of them yields an alternative that contains only the characteristics of the given theory (which is implied by the one under analysis) and yet imply a relation that is not contained in that given theory, this relation is synthetic *a priori* knowledge. As the infinity of theories is defined by anti-implication of the given theory, any one of them can yield an alternative through deduction (implication) that contains only the characteristics of the given theory merely by stating that the given theory is implied by the one under analysis. This alternative, however, is only the given theory itself, which naturally contains no other relations than was in it to begin with, and so it is not synthetic *a priori* knowledge. To synthesize a new theory from a given one so that the new theory contains only the characteristics of the given one (and thus has the same extension) but in a new and synthetic relation, one of the infinity of theories that imply the given theory must have additional definitions that are productive.

They must be so constituted that, in the analysis of that theory, these additional definitions are conjoined with the given ones in such a way that the characteristics re-form into only those of the given definitions (the superfluous ones being dropped by deduction) but in a different relation than they were in when arranged in the definitions of the given theory. Conversely, if the analysis of an anti-implication yields an alternative that contains only the relations of a given theory and yet imply a characteristic that is not contained in that given theory, this characteristic is called synthetic *a priori* knowledge.

Unfortunately, there is no algorithmic method for finding such a theory out of the infinity of anti-implications of a given theory. What superficially seems to be the most difficult part of scientific research, thinking of hypothesis, is the easiest because there is an infinity of them. The difficult part is recognizing the one that can have its superfluous characteristics analyzed out of it and still leave synthetic *a priori* relations. This is where quality enters into science. An anti-implication that can yield synthetic *a priori* knowledge is productive. It conforms to a definition at a high place on one's value scale, and that is how it is recognized. Successful recognition of an anti-implication that yields synthetic *a priori* knowledge is hampered by a subject-object dualism which makes a researcher see only the infinity of possibilities, while whoever is at one with his work goes directly to a solution. A researcher with a dualistic attitude about his work goes through the same motions of searching but does not see what he looks at.

The use of additional definitions which are then deducted after a solution has been found is often forgotten, leading people to believe that synthetic *a priori* knowledge is impossible and that all understanding is analytic. That synthesis is a passing event which leaves no mark on its creation and that all declarative sentences are analyzable from discursive postulates has led many linguists to take this stand. As linguists deal with theories whose creation has been forgotten and which have turned into statements that could have as easily been handed down from a mountain as synthesized, it is not surprising that they should regard these as cases of analytic knowledge. They need only ask "What do the words mean in this configuration?" and they know the meaning of the theory. They forget that at one time the theory was unknown but a simpler one was known without certain relations. Then people noticed that, whenever they used the theory, the phenomena that conformed to it had those relations, but they were hesitant to risk anything on the assumption that future phenomena would have those relations also, for they could not be sure that it was not a coincidence. Then someone found an anti-implication of the theory which, when analyzed, yielded those relations as synthetic *a priori* knowledge, so people no longer had to wonder if the next phenomenon that conformed to their theory would have those relations but could relax and say "Whatever phenomena conform to these characteristics has these relations." Or it might have happened another way and it was some of the characteristics which people were unsure of and someone found an anti-implication of the known relations that, when analyzed, yielded those characteristics as synthetic *a priori* knowledge. But that has all been forgotten and now linguists only see a theory with certain characteristics and certain relations, so they analyze it and then proclaim that synthetic *a priori* knowledge is impossible. An illustration of synthetic *a priori* knowledge will now be given:

If definition *p* is of a given equilateral triangle XYZ and definition *q* is of a square, what additional definition(s) *r* must be added to *pq* so that the analysis of *pqr* leaves only the characteristics of *p* and *q* (there are no superfluous lines) but with the square given definite size, ABCD, so that it fits inside triangle XYZ, as shown in figure 6?

Figure 6

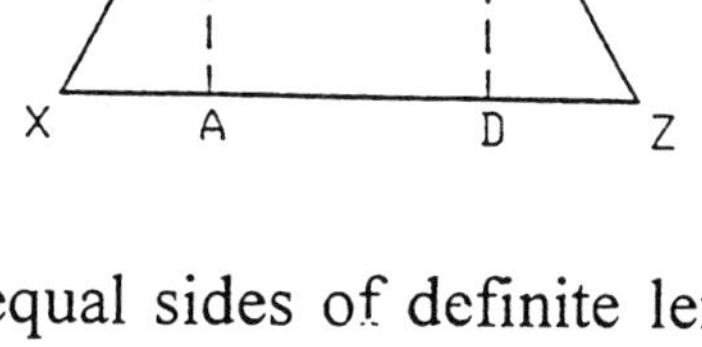

Neither the definition of *p* (three equal sides of definite length) nor the definition of *q* (four equal sides of indeterminate length) contains any information about the position of B or C. Additional characteristics are needed (lines have to be drawn) to find these lengths. But these additional characteristics have to be deducted again for the new theory to keep the same extension. It is easy to delude oneself in this sort of exercise by assuming that the additional definitions needed are contained in the "full" meaning of the given definitions, even though they are not strictly expressed in them. It is the mark of an immature science to rely on such assumed meaning; only when a science defines its terms in the strictest way possible can it truly be called a science.

Figure 7 illustrates the solution. The first additional definition needed is of another equilateral triangle with XZ as one side. Next, connect the opposite vertices so that the line crosses XZ at its midpoint. Then, construct a square with XZ as one side (additional triangles must be constructed to prove the sides of the square perpendicular to XZ, but they are not shown). Finally, draw lines from the midpoint of XZ to the upper corners of the new square. The points where they cross XY and YZ are the definitive positions of B and C. Connect B and C and drop perpendiculars from them to line XZ to find points A and D of the square inside the equilateral triangle. This figure, *pqr*, is an anti-implication of *pq*, as it can imply *pq* by deducting the additional lines just added. They are not needed for phenomena to conform to the square inside the triangle. After their deduction, however, the relation of B and C to lines XY and YZ, which was not known before, is still there. This relation is synthetic *a priori* knowledge.

Figure 7

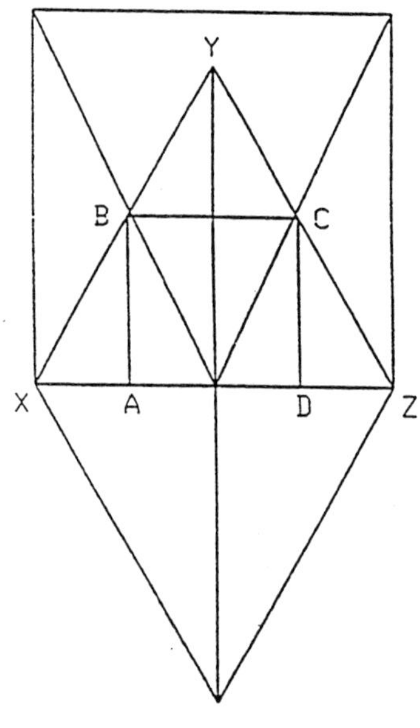

A more algebraic example is the integration of $\frac{1}{\ln(x)}$. The first three steps establish the needed anti-implication.

$$\int \left[\frac{1}{x\ln(x)} + \frac{x}{x\ln(x)} - \frac{1}{x\ln(x)}\right] dx$$

Multiply by $\frac{x}{x}$, add and subtract $\frac{1}{x\ln(x)}$.

$$\int \left[\frac{1}{x\ln(x)} + \frac{e^{\ln(x)}-1}{x\ln(x)}\right] dx$$

Substitute $e^{\ln(x)}$ for x in numerator.

$$\int \left[\frac{1}{x\ln(x)} + \frac{1}{x\ln(x)} \sum_{n=1}^{\infty} \frac{\ln^n(x)}{n!}\right] dx$$

Substitute for the numerator its Taylor series expansion.

$$\int \frac{dx}{x\ln(x)} + \int \sum_{n=1}^{\infty} \frac{\ln^{n-1}(x)}{xn!}\, dx$$

Separate and bring $\frac{1}{x\ln(x)}$ into the summation.

$$\int \frac{du}{u} + \int \sum_{n=1}^{\infty} \frac{u^{n-1}}{n!}\, du$$

Substitute $u = \ln(x)$ and $du = \frac{dx}{x}$.

$$\ln(\ln(x)) + \sum_{n=1}^{\infty} \frac{\ln^n(x)}{nn!} + c$$

Anti-differentiate and put $\ln(x)$ back.

Mathematicians will confirm that much of what they do involves finding a "trick", usually an identity, which transforms a given equation into one which superficially seems more complicated but which in fact easily implies what they are trying to prove. Almost invariably, once this trick is discovered, the rest of a proof is just a matter of algebraic manipulation.

5.3 General theories and universality

All phenomena conform to definitions on the transcendental aesthetic, so their conformance is a definition to which they all conform. There may be other definitions, however, to which all phenomena conform. A definition to which all phenomena conform has the characteristic of universality. This characteristic is only implied by the definition of conformance to definitions on the transcendental aesthetic; with any other definition, its having that characteristic is a conjecture. But some theories have such universal application that it is easy to imagine that all phenomena conform to them. Such theories are called general theories and their existence is credited to divine teleology.

The conjecture of there being purpose in noumena (a Creator) arises from the existence of general theories not implied by the definition of the transcendental aesthetic. The conservation laws of physics are not sufficient, for they can be implied by the symmetry of the transcendental aesthetic. For instance[6], if an object isolated from external forces begins to rotate, it has to do so about an axis, but the choice of that axis selects a direction in space, which would otherwise be symmetrical. Hence, the Law of Conservation of Angular Momentum is implied by the spatial dimensions having no natural orientation. Similarly for the Law of Conservation of Linear Momentum, because, without violating the Law of Conservation of Energy, the object discussed above could only acquire its spin by exchanging linear momentum for angular momentum. But the Law of Conservation of Energy is implied by the symmetry of the temporal dimension. Suppose that gravity were systematically weaker on certain days. Then one could pump water uphill on those days and let it generate more energy by flowing down on the other days, in violation of the Law of Conservation of Energy and of the symmetry of time.

But there may be other, equally general laws for which no anti-implication can be found that, when added to the definition of the transcendental aesthetic, imply that law but which can still be deducted so as not to import any new meaning to the five transcendental dimensions. Possible candidates for this anti-implication are infinite, however, so it is impossible to prove the existence of the Creator on the basis of one's failure to synthesize an apparently general theory from the definition of the transcendental aesthetic. Yet, ultimately, anyone who believes in a Creator can trace his belief back to a general theory which seems to require "something more" than the transcendental aesthetic for its universality. Arguments such as those presented in the first chapter of this book may strip that person's beliefs of their institutional trappings and their anthropomorphic caricatures. He may even quit praying for guidance and donating money to evangelists, but his basic belief cannot be shaken.

Because the Creator is thought of only in conjunction with divine teleology, It cannot be regarded as synonymous with noumena but only as a part of noumena, for It could only have created those laws that were discovered as general theories out of *something*. This noumenal *something* is the opponent which the Creator had to overcome in order to display Its omniscience in such a way that people might think they see it in general theories. As far as the Creator is thought of in connection with divine teleology (and there is no other way to think of It), It is conceived of only as bringing order to noumena and not as the material of noumena, this function belonging to that *something*. If this is true, general theories that have held without exception might yet break down in observation of the extremely small - the very matrix of all creation. For, while the Creator has erected Its edifice out of matter and energy, the nature of this matrix is of that *something*. That *something*, however, is also a noumenon and is thus as unknown to people as the Creator is. It is a confusion to think that people can break through the *limit* which separates phenomena from noumena merely by changing the scale of their observations. The fact that atomic particles are known at all implies that they are not noumena. The fact that they do not have position in the usual sense of always being at a definite place implies that they are not phenomena either. In a sense, nothing is, as the spreading out of position is not limited only to atomic particles, it just becomes noticeable at that level. Thus, the definition of phenomena, like most definitions, requires some tolerance to admit any application. Atomic physics requires more tolerance than other sciences and, hence, the transcendental aesthetic is less useful to atomic physicists than it is to other scientists. This just means that books about atomic physics have fewer pictures. The transcendental aesthetic is not the only tool available to scientists, though phenomena must have a position, however spread out, to be known at all. To have position is to be real, and scientists in all fields concern themselves only with what is real, not imaginary.

5.4 Extension in the three senses of truth

The extension of a theory in each of its three senses of truth are as follows:

extension	none	some	all
first-sense	unobserved	observed	universal
second-sense	contingent	applicable	general
third-sense	inconsistent	consistent	tautological

If no phenomena conform to a definition, either in all of its characteristics or all of its relations, phenomena are unobserved and that definition does not have first-sense truth. If some phenomena conform, they are observed and the definition is true in those cases. If no phenomena have ever been observed that did not conform, that definition is not only true but might have the characteristic of universality.

If no phenomena conform to all the characteristics or all the relations of a theory, that theory's application is contingent on such conformance. Whenever phenomena conform, that theory is applicable and may predict the relations if it is to the characteristics that the phenomena conform, or the characteristics if it is to the relations that the phenomena conform. If all phenomena that have ever been observed have conformed to all the characteristics or all the relations of a theory, its predictions might be general.

If no phenomena could possibly conform to a theory because all of its 2^n alternatives contain contradictions, that theory is inconsistent. If some of those alternatives could conceivably have phenomena conform to them (even if none has), that theory is consistent. If phenomena cannot do anything but conform to a theory (it takes phenomena any way they come to it), that theory is tautological.

CHAPTER III

RATIONALITY

A complete knowledge on the part of an observer of external influences might allow prediction of the form to be taken by action only because the logic of the observer enables him to know with certainty the path that the actor's own logic will select.

Israel Kirzner
The Economic Point of View[7]

Section 6

6.1 Empirical observation in economic theories

Errors in first-sense truth are found in the application of all theories to a greater or lesser degree. Phenomena conform to definitions and, because we are not omnipotent, we cannot make phenomena conform perfectly but can only approximate such accuracy. Whether this conformance can be made accurate enough for practical purposes depends on the situation as well as on the importance of such precision. While this has no effect on a theory's truth in the third sense, it can prevent a theory from being of much practical use even in situations to which it clearly applies. This is especially the case of theories that are dependent on knowledge of the importance of phenomena to other people. This importance is never given directly but can only be known as greater or less than the importance of other phenomena (the money needed to make a transaction, for instance). Nothing else is so jealously guarded as one's value scale. Indeed, one does not know it fully until one is called on to make a decision and, if the decision comes easily, one may still not be clear about one's values. The importance of phenomena varies considerably in fairly short periods of time, further compounding the difficulty when an action must be taken in anticipation of certain phenomena being important to people in the future. Because of this, an economic theory that is deduced from axioms cannot extend the implications of its axioms with statistical inferences to predict future events. As long as the predictions of the theory are deduced only from universal assertions about all human behavior, the theory is on an arguable footing. But, when a theory branches off so that different predictions are made in different empirical situations, everybody's statistics indicate a different situation and it is impossible to argue for or against the truth of the theory.

However, while refraining from extending the implications of one's axioms with statistical inferences can keep one's theory on an arguable footing, it is still arguable; that is, to establish second-sense and third-sense truth, one must still provide intuitive reasons for one's choice of axioms and show that there are no gaps in the deduction of one's implications, respectively. For theories which one intends to apply, establishing second-sense truth is more difficult than establishing third-sense truth and, to the extent that it fails, the predicted results may be greater or less than expected or even entirely offset by the effects of other actions not considered by the immediate theory. However, as far as one believes in the applicability of one's axioms, one knows with certainty whether one's actions aid or hinder one's purposes to the extent that phenomena are affected by them. If one's theory is not deduced from axioms, however, one has no way of knowing whether observed phenomena are caused by one's actions or occur in spite of them. If one's theory cannot be proven with logic alone, one must rely on *post hoc ergo prompter hoc* reasoning which bases one's decision on what was observed to have followed similar actions in the past. If one likes the phenomena that followed those actions, one may believe one's opinion proven even though one's critics may produce a dozen examples where it happened to be followed by negative observations, and in turn one may produce a dozen more to support one's opinion. Such anecdotal arguments in which each belligerent attempts to outdo the other with stories of similar situations which illustrate his point are idle, for they prove nothing. In fact, immediate results are usually opposed to the overall effects an action has on the attainment of quality.

Only an axiomatic economic theory, abstract from all experience, can prove with certainty whether an action helps or hinders a given purpose. With such a theory, one can be content that one is doing the best one can even if one's opponents can produce statistics showing that the immediate results are negative. Similarly, any action that such a theory predicts will have effects greatly detrimental to quality in the future, despite a small immediate attainment of quality, may be abstained from with certainty despite an immediate lack of examples demonstrating its negative effects on quality, which only appear in the long run. Economic theories take the form: "If a certain action is taken it will have negative effects now but greater positive effects later", or conversely; "If a certain action is taken it will have positive effects now but greater negative effects later." They are amoral; that is, they do not make any judgment as to whether an action should be taken but merely state the relation between an action and its effects. But people must be convinced of this relation on purely logical grounds. If people accept arguments by example, everybody will have their own statistics to cite and nobody will know what to believe.

To make business management decisions regarding specific phenomena, however, one must extend the implications of one's axioms with statistical inferences. While the assertions of economic theory are deduced only from axioms without reference to the parameters of a theory which distinguish different phenomena from each other, business managers must fix these parameters with regard to the specific phenomena in which they deal. In general, this cannot be done very accurately, which is why the preceding paragraphs have emphasized the importance of abstract economic theory disassociating itself from statistical inferences. For abstract economic theory to gain the respect of the hard sciences, it must avoid statistics and make assertions based only on its axioms, that is, assertions which are true regardless of how the parameters are fixed which distinguish different phenomena from each other. Business managers must fix the parameters describing their phenomena as accurately as possible, however, and for that they must use statistics.

Economic theory tells one the best action one can take in a certain situation. If one knows exactly what that situation is (one can accurately fix one's parameters), economic theory tells one the perfect action to take for phenomena to conform to the highest possible definition on one's value scale. The problem is acquiring the empirical data about a situation, as this data consists of knowledge of other people's value scales, which is difficult to obtain with accuracy. One may be mistaken as to the extent that people value certain phenomena, in which case economic theory will predict the best action to take in a situation that does not actually exist. When one applies this action to the situation that does exist, one will be thwarted by counter-forces that one did not know about or whose relevance one underestimated. Economic theory is abstract and, if it is proven true in the third sense, it is neither responsible for nor refuted by the failures of those who use it, their failures being caused only by a lack of accurate empirical data. Unfortunately, there is no algorithmic method for obtaining such data, especially for obtaining it about the future, which is where it counts the most, as all actions require time for completion after one has committed oneself to them. To be successful, however, one does not need to be perfectly informed about a situation, only better informed than other people. This is why business management can only be considered an art and not a science. Abstract economic theory can be regarded as a science as long as its assertions stop short of requiring the parameters of the theory to be fixed in any particular way.

In most problems to which an economic theory is to be applied, that theory defines all the factors that could affect the user's attainment of quality, so one need not fear the appearance of completely unknown counterforces. The question is rather: What is the relevance of each of these factors? How much influence each definition in an alternative has on that alternative's position on one's value scale is usually known in a limited, intuitive way, but formal logic has nothing quantitative to say about the relevance of definitions conjoined in an alternative. Human actions cannot be predicted with statistics the way physical actions can. The concept of probability in the social sciences is intuitive only and does not justify numerical techniques. Unfortunately, many people are impressed by the appearance of mathematics even when they do not understand its use, and it is for this reason that statistics are applied so often in the social sciences. If economists are ever going to gain the respect of mathematicians, they must make only those assertions that can be deduced from the axioms of their theory and refrain from extending it with statistical inferences.

6.2 Empirical observation in ethical theories

When there is a large probability that phenomena conform to certain definitions, yet their conformance is difficult to determine in specific situations early enough to be of use, certainty is often attached to their conformance. This is especially the case of theories concerning actions whose failure is dependent on the conformance of phenomena that, by its nature, remains concealed for a period of time after the action is initiated. Such theories are called ethical theories. Ethical theories take the form: "If a certain action is taken it will have negative effects now but, if a second phenomenon appears later (though concealed now), the action will have greater positive effects later", or conversely; "If a certain action is taken it will have positive effects now but, if a second phenomenon appears later (though concealed now), the action will have greater negative effects later." If phenomena conform to the first definition (the action is taken), phenomena are assumed to conform to the second definition even though it cannot be verified until after the action is initiated. This assumption, made before verification of first-sense truth is possible, is based on the probability of phenomena conforming to the second definition whenever phenomena conforms to the first definition. It is accurate to the extent that phenomena conforming to the second definition is implied by phenomena conforming to the first definition. If it is wholly implied by phenomena conforming to the first definition, the two definitions are combined into one, which phenomena either conform to or not. In that case, the theory is an economic theory and not an ethical theory.

What separates ethical theories from economic theories is that, in the former, the conformance of phenomena to the second definition is not always implied by phenomena conforming to the first definition but only implied when another factor is conjoined with that first definition. That other factor is whether an individual is discovered taking the action that conforms to the first definition. To state that an ethical theory is true is to say that the probability of one being detected whenever one takes the specified action is great enough to be considered certain. While it is always possible to point out examples where actions are taken that conform to the first definition and phenomena do not conform to the second definition, most people will realize after reflection that this assumption is not unfounded in most cases.

Section 7

7.1 Reinforcement of generally accepted ethical theories

To be considered an ethical theory, the probability of phenomena conforming to the second definition whenever the first is met must be high enough to be considered certain, so the acceptance of an ethical theory by society is dependent on this probability. At the same time, however, the general acceptance of an ethical theory by society has an effect on that probability, so whether or not one accepts that probability as high enough to be considered certain is somewhat dependent on whether other members of society have already accepted it as certain. If they have, the certainty of that ethical theory is strengthened by their efforts to see that phenomena conform to the second definition and that they do so conspicuously for the example of other people. Whenever the second definition of an ethical theory negates the good effects of the first definition and results in an overall negative effect on quality, one finds more people than would otherwise be involved taking actions (punishments) that conform to this second definition and to a greater degree. This causes the negative effects to be greater than they would have been if the truth of the ethical theory were not generally accepted by society. Similarly, whenever the second definition results in an overall positive effect on quality, one finds strangers taking actions (rewards) that conform to this second definition, making the benefits of taking the actions that conform to the first definition greater than they might have been otherwise. Failure of society to recognize the truth of an ethical theory has the opposite effect by decreasing the probability of phenomena conforming to the second definition whenever an action is taken that conforms to the first definition. Because of a misguided morality, people may abstain from taking actions that conform to the second definition (either punishments for bad actions or rewards for good actions) which they otherwise would take if their mistaken conception of ethical theory did not enter into their decision.

The effect of this relation between the truth of ethical theories and their acceptance is to make it profitable for one to conform one's actions to those defined as good in the ethical theories accepted by society and to abstain from those defined as bad, whatever one's personal understanding of the truth or falsity of any particular ethical theory may be. This often causes doubt as to the truth of ethical theories in general because, while their truth is dependent on inference, the example of those who possess a correct understanding and base their actions on that understanding is often poor when they do so against the common consensus. Because the effect of an ethical theory's acceptance on its truth is forgotten and only the effect of its truth on its acceptance is considered, people who follow ethical theories whose truth is clearly seen, though not by society in general, not only fail but are often ridiculed. This may cause an observer to believe that all ethical theories are fallacious and the method of finding their truth is flawed. For this reason, errors of society concerning the truth of ethical theories cannot often be corrected by the example of the success of those who have a better understanding and follow their own conscience against the dictates of society. Errors are only eliminated through the gradual increase in understanding of society, so the first people who possess this understanding usually do not profit from their knowledge but can only go along with the existing theories.

7.2 Basis for the acceptance of ethical theories

Whether or not one acts according to an ethical theory is dependent not only on one believing the relation predicted by the ethical theory, but also on one caring what the future effects of one's actions will be. An individual, blinded by the immediate quality an action brings him, may not see that the further effects of his action bring more damage to his attainment of quality so that, in the long run, he is better off not to take that action. Most crimes are committed because people do not care about the future and not because they do not think they will get caught. This is why so many transactions turn into swindles even though the perpetrator knows that as soon as his victim examines what he has received, he will realize that he has been swindled. The perpetrator cannot see far enough into the future to anticipate the victim coming back to retaliate.

The science of ethics is primarily concerned with promoting the long-term interests of people over their short-term interests, so to speak of such things as "society's interests" or "duty for duty's sake" is a mistake. Nobody gets very excited about the interests of society; they are looking out for themselves. Those interests traditionally ascribed to society are generally so close to the long-term interests of an individual that they can be considered as one. If this is so, they should be called by their proper name so that people will see ethics as a program they can support, and not a liability. Attempts to prove the truth of ethical theories without reference to individual quality are not only futile but detrimental for, if they convince students of anything, they convince them that ethical rules are opposed to their well-being and that they can do nothing better for themselves than ignore them. Whatever moralists say, nothing drives people to action but their own attainment of quality. It is not spirits which move people to action, only quality. Talk of "animal spirits" is one of the more ridiculous chapters in economic history.

To say that one should only take actions that would benefit society if they were adopted as an established guideline which everyone follows is unconvincing, for one knows that one's single action among all the countless actions taken by the rest of society will not set a precedent. Little good has ever come from those who affected to promote the common good through their actions. Only when people think of their own welfare does the wealth of society increase as though by a divine plan. One who arrogates to himself the knowledge of what is good and what is not good for society is erecting himself as God. One's beliefs as to what is good for society are dependent on nothing but what one deems desirable of becoming an established guideline. There is no other method of knowing what is desirable as an established moral guideline except by imagining the resulting situation's position on one's own value scale. No moral guideline is categorical; some are just more desirable than others.

The judgment that rational nature exists as an end in itself is also unconvincing for, as all actions are purposeful, all actions are rational.

Only individual, sentient beings experience quality. Rational nature is the definition to which they conform as phenomena. As the definition of rational nature does not imply that those phenomena conforming to it have similar purposes but only that they are purposeful, there is no way to direct one's actions to fulfill any of these purposes without offending others. What is demanded by ethical theories is not that one treat other people as ends in themselves but only that, when treating one's own attainment of quality as an end, one should regard other people in their long-term use as means only, and not be swayed by the immediate advantage acquired by their exploitation. One should rather realize that such crass actions are so often followed by greater negative effects that the phenomena on which these effects depend may be considered certain whenever the first phenomena (the action) occurs. With this inference, one will only use other people as far as they cooperate voluntarily so that one may be assured of their future cooperation, which is so essential to one's overall attainment of quality. Social cooperation and the general attainment of quality is not based on mutual love but rather on everyone's prudent use of other people as means to their own ends.

Section 8

8.1 Difficulties in the application of economics and ethics

Economic and ethical theories are not denied completeness any more than the physical sciences. A theoretical system can be built to which all pertinent phenomena can conform so that one is never confronted with an event that defies explanation. This is true of any science that concerns itself only with phenomena. Only a science to whose definitions other definitions may conform, according to what intensions they have, is incomplete. Set theory is an example of this because sets may be members of other sets; so is linguistics because words are distinguished by their conformance to other words. But economics and ethics have problems of their own. They are plagued by a lack of reliable observations of phenomena on whose conformance its predictions are contingent. To make accurate predictions, one needs to know the importance of phenomena to other people. This data is not available. Psychologists may arrogate to themselves that they know these things about other people, but that is vanity. All that can really be known about a person's value scale is that, at a particular time, he chooses one phenomenon over another. Whether the choice comes easily to him or is made only after great reflection cannot be seen. Furthermore, the importance of phenomena varies considerably in time and is affected by phenomena having conformed in the past. Thus, having observed someone acquire a sample of phenomenon implies that the importance of that phenomenon has changed for him. Asking a person is out of the question for he will assume that the questioner is interested in conducting a transaction and he will conceal his value scale in the hopes of bettering the bargain. If he has no interest in dealing in that phenomenon, he will be annoyed by the question and either refuse to answer or tell a lie. Also, there are too many people who affect a market to try to observe them all. The quantity of information needed for determining the perfect action in every situation would be staggering even if it could be had.

If a phenomenon is fungible (interchangable), the number of units that have already conformed to a definition affects the importance of the one in question as an individual grows sated. The number expected to conform in the future has a similar effect and, furthermore, where one is only expecting phenomena to conform, the probability of that conformance must also be taken into account. This prevents the phenomenon in question from being invariant from one situation to the next, as its importance at any instant is dependent on other phenomena not contained in its definition. This is called diminishing utility ("utility" is synonymous with "value"). It will be discussed in more detail in the next chapter.

The value of a phenomenon may be affected by the occurrence of not only exactly similar phenomena but also phenomena similar enough to act as substitutes. Their conformance increments the phenomenon's position on one's value scale, not by a full occurrence, but by a fractional occurrence. This makes diminishing utility continuous. Therefore, importance is only stable when a phenomenon is permanent but of such a nature that one never becomes sated with it or its substitutes. Money conforms to this definition and can be used as a measure of utility.

Some economists have disputed the assertion that money is stable and, indeed, Chapter Five introduces institutions which can have a dramatic effect on the value of money. The motivation for this dispute has often been the mistaken view that the stability of even one phenomenon is sufficient to prove the cardinality of value. It is not. All phenomena would have to be independent of one another's conformance for one's value scale to be cardinal. Motivation aside, however, the assertion that the value of money is derived historically (see section 10.1) is sufficient to prove that money is stable, for its value is uniquely a function of speculation on the effects of the above-mentioned institutions as well as on slower-moving parameters such as the assimilation of isolated communities. If the causes of fluctuations in value of a phenomenon exceed the comprehension of speculators, that phenomenon is neither stable nor does it function as money and, hence, it is not a contradiction to identify money with a measure of utility. The existence of price-setting implies the stability of money, though the institutions discussed in Chapter Five cause the purely speculative part of prices (the price-change premium) to take on more and more importance, especially in credit transactions, until people revert to another form of money. But at every point in time, money is a measure of utility because of its stability.

The stability of money does not imply that one's value scale is cardinal; all phenomena would have to be stable (independent of phenomena's conformance to other definitions) for that to be true. Most phenomena are very much dependent on the previous or expected conformance of other similar or complimentary phenomenon, which decreases or increases their value, respectively. A sign of the instability of a phenomenon is how often it seems to indicate intransitivity in one's value scale, for one's value scale is, by our first axiom (see section 11.1), transitive. All apparent intransitivities are caused by phenomena whose values change from one application to the next. Let us say that a subject is offered a phenomenon that conforms to definition p or q, but not both, and he chooses p. Then, when offered a phenomenon that conforms to q or r, but not both, he chooses q. Finally, the subject is offered a phenomenon that conforms to p or r, but not both. The subject chooses r because he has already had p and is sated on that kind of phenomenon and has decided to get one of each sample. The researcher believes that he has demonstrated the existence of intransitivity. However, it is simply that, for this subject, $pqr > pqp$. p is unstable because the subject becomes sated on them so fast. Neither does it work for a researcher to offer all the p, q, or r in existence because utility does not necessarily diminish uniformly for each of these definitions, so the relative importance of these masses of phenomena do not correspond to the relative importance of individual units. In any event, a researcher does not have all of any of these phenomena to offer nor can a subject imagine actually getting them, so the question is hypothetical. When given hypothetical questions, people are inclined to lie. More important to the subjects of the research than anything a researcher can afford to offer, however, is the sport of confusing the researcher with arbitrary answers to his questions. It must be remembered in all surveys that what is being measured is not what the survey purports to study but rather what people tend to tell poll takers. Thus, surveys do not measure what they intend to measure, so there is no reason to think that their results are normally, or otherwise, distributed around the true value.

8.2 Ignorance of the intellect and freedom of the will

While economics and ethics have fundamental problems with their data that prevent them from making accurate predictions of human action, technology is also less than optimal. Perfection being defined as "exactly fitting the need in a certain situation or purpose", human actions can never be perfect because of problems in applying the appropriate theories. Only if one were omniscient could one's actions be considered strictly predetermined and thus perfect. As everybody is ignorant to some extent, their actions can never be wholly predetermined but rather are taken *ad libitum* to a large degree. One must make the best of one's limited knowledge in choosing the course of action that seems most perfect. One can never know which action is perfect or even, to an uncomfortably large extent, which is most nearly perfect. Any choice made *ad libitum* is autonomous and cannot be predetermined. No hidden phenomenal forces can affect one's decisions if one is ignorant of them. Ignorance is what makes life interesting. If people were omniscient, there would be no excitement in finding a clever method, for one would already know the optimal one and would perform it mechanically. Such mechanical actions would be predetermined, for any other omniscient individual could duplicate the reasoning of his colleagues to predict their actions. Ours, however, is a world of ignorance and even the most astute individual cannot predict half the strange plans people conceive. Thus it is that we have free will.

8.3 The definition of the term "rational"

All human actions are rational, as "rational" is defined as any action taken to satisfy one's purposes or wants. More concretely, rational actions are those directed at creating wealth, though in a much broader sense of the term "wealth" than it is taken by accountants. (Notice that the allocation of scarce resources is only incidentally implied by this definition.) Rationality does not imply only long-term and not short-term interests for, if it did, no action would be rational for no action could consider infinitely long-term interests. Nor does rationality imply any level of competence in designing one's actions, for all actions are taken with a certain amount of incompetence. If an action fails to attain its purposes, one is inept, and if those are only immediate purposes and one is later sorry one took that action, one is shortsighted, but this has nothing to do with rationality. If one wants to commend an individual for a well-thought-out course of action that he has not regretted, even long after it has passed, there are perfectly good words for that, but rationality is something different. It is what distinguishes human actions from natural events.

Economics and ethics are concerned with human actions because they are rational. Economic theories are those that make predictions which concern human actions whenever phenomena conform to them. Ethical theories are those that make predictions which concern human actions whenever certain phenomena conform to them immediately and other phenomena conform to them later but, because the latter phenomena appear too late for the prediction to be used, they are assumed to conform if the first phenomena did. Together, these two sciences are what guide people in their actions.

PART TWO

ECONOMIC THEORY

You show the world as a complete unbroken chain, an eternal chain linked together by cause and effect. Never has it been presented so clearly, never has it been so irrefutably demonstrated.... Whether it is good or evil, whether life in itself is pain or pleasure, whether it is uncertain - that it may perhaps be this is not important... But according to your teachings, this unity and logical consequence of all things is broken in one place. Through a small gap there streams into the world of unity something strange, something new, something that was not there before and that cannot be demonstrated or proved: that is your doctrine of rising above the world, of salvation.

Hermann Hesse
Siddhartha[8]

CHAPTER IV

PRICE AND STOCK

Needs arose from our drives and the drives are imbedded in our nature. An imperfect satisfaction of needs leads to the stunting of our nature... but to satisfy our needs is to live and prosper.

Carl Menger
Principles of Economics[9]

Section 9

9.1 The definition of the term "marginal utility"

Definitions to which one or more phenomena may conform do not exist at one point on one's value scale but rather in a series of points labeled "1st occurrence", "2nd occurrence",... The intensions of the definitions are the same at each of these points, the importance of each position being different because of factors not contained in the definitions, that is, how many phenomena have come before or are expected to come. The spacing of the definitions in a series is not even but is determined by diminishing utility. Phenomena that conform to the definitions in such a series are fungible, meaning interchangable. Being interchangable, they cannot each have a different value (importance), for the loss of one being employed for an important purpose can be met simply by replacing it with the one that conforms to the definition of marginal utility. Because any of the phenomena conforming to definitions in a series can be replaced by the one with the least utility of those being satisfied, one does not value any of them more than the last one. Marginal utility is all that is ever at stake when risking a unit of fungible phenomena. When considering the acquisition of another unit of fungible phenomena, the value of that unit is the utility of the next want to be satisfied in its series. In either case, the value of a phenomenon is never determined by the use to which it happens to be applied but by the use on the margin between satisfaction and nonsatisfaction; hence the term "marginal utility."

The smallest bit of phenomena which still conforms to a definition (satisfies a want) is the stime of the definition. Some definitions have a very definite stime which cannot be broken into parts, so whatever unit one chooses to deal with must be a multiple of this basic unit. Others, however, can be broken down almost indefinitely without losing their meaning, so the choice of the unit of these phenomena is more arbitrary. Either way, one chooses the size of a unit according to its convenience in transactions. Different-sized units are appropriate for different transactions. The spacing between the definitions in a series is not necessarily even, so the marginal utility of one size unit is not necessarily proportional to the marginal utility of another size unit.

If the unit that is chosen encompasses all the phenomena that could conform to a definition, it is of academic interest only, for it is very rare that choices have to be made concerning such a large unit. The value of such a unit is its marginal utility but, as the entire field is (theoretically) already acquired, its value is the sum of the utilities of all the definitions in the series out to infinity, assuming that it converges. It is not surprising that phenomena that conform to series of definitions that, taken all together, are very important (some of which, if no phenomena of that kind existed, would cause the immediate extinction of mankind) are fortunately of such abundance that no reasonable unit has any meaning at all. Other phenomena that are considered luxuries are nevertheless very important even in small units because of their scarcity. If the total quantity of the former were artificially restricted, however, the value of any substantial unit of them would exceed the value of a unit of the latter, for, while they were both scarce, the former would be a necessity while the latter would remain a luxury. The measure of the quantity of a phenomenon is only meaningful relative to the requirement for that phenomenon. The requirement for a phenomenon is the quantity needed to fully satisfy the want to which it conforms, that is, to bring its marginal utility down to (virtually) zero. Fungible phenomena of such abundance that no reasonable unit is ever greater than their excess over the requirement for them are free. Their marginal utility is zero, so they are of no concern to economic theory. Economic theory is concerned only with phenomena of such scarcity that their quantity is less than or equal to the requirement for them and which are thus important. Phenomena of such an abundance that they are a nuisance are not dealt with directly, but rather it is the service of removing them that is traded. Thus it is that everything has a positive price.

9.2 The definitions of "use" and "exchange" value

It is sometimes useful to imagine a definition with certain characteristics deducted and then to determine its new position on one's value scale. One instance of this is the deduction of exchangeability because, despite people's voiced intentions or the fact that they have bought similar phenomena in the past and appear to need more, one cannot be sure that they will buy any until a transaction is complete and then the question has lost its relevance. The importance of phenomena conforming to a definition with the characteristic of exchangeability deducted is their use value. It is the importance that one associates with one's personal use of a phenomenon.

Another instance of deduction is of one's ability to use the phenomena that conform to a definition. This deduction can be accomplished by imagining that the phenomena no longer have any use or that one is prevented from using them, in which case one will immediately put them up for sale, accepting whatever is offered for them in the time span one can allow oneself. Out of the proceeds is subtracted the cost of advertisement and storage to get exchange value. Use value is never uncertain as it depends entirely on one's own value scale, but it may be vague as far as one does not know one's own will. Exchange value, however, is uncertain because it depends on the actions of another person (the buyer) and it is for this reason that use and exchange value are calculated separately; in certain situations exchange value is ignored because of its uncertainty.

9.3 The effect of exchange value on marginal utility

When defining marginal utility in section 9.1, the quantity of phenomena conforming to a definition was considered to be a constant of which value was a function. Where units of a phenomenon can be bought and sold and more of them produced out of the necessary labor and capital, the quantity of phenomena is variable and it must be shown to be a function of a constant lest two variables be defined with one equation.

The quantity of phenomena conforming to any particular definition cannot be entirely variable, for any addition to it entails foregoing the acquisition of phenomena in another series of definitions. Neither are any units sold unless they bring something of greater value. Whenever the exchange value of a phenomenon is greater than its use value, units of that phenomenon are sold until exchange value and use value are equal. Exchange value drops because one has become sated on what is being received in the exchange and use value rises to meet it as units are lost in exchange, making the remaining ones more valuable. This is true of all phenomena which one possesses, including one's own labor. That it is true of labor is obvious to an employee who hires himself out for certain hours of the day during which time he is expected to devote his energies to the furtherance of his employer's purposes but can do as he pleases during his off hours. It is equally true of a businessman who is his own master, however, as some of his labor is directly useful to him, particularly leisure activities, and some is expended in the manufacture of other phenomena that are then valued for the definition to which they conform.

Because one's wealth at any time is constant, it is applied first to the high end of one's value scale, producing phenomena or exchanging phenomena already acquired for those which conform to definitions at the top of one's value scale and continuing down until all of one's wealth is exhausted. In this finite quantity of definitions with phenomena conforming to them, there are a fixed number of definitions in each series of similar definitions which have phenomena conforming to them. Marginal utility is defined with the quantity of phenomena conforming to each definition, not an arbitrarily fixed constant, but a function of wealth.

9.4 The relation between wealth and happiness

The wealth of a person (or community) is his (or their) stock of phenomena. Because people work on the high end of their value scales first, it is convenient to think of wealth as a pointer on one's value scale, above which one possesses phenomena in each series and below which one does not. Poor people can only satisfy their most important wants, so their pointer is very high on their value scale. Only the first few terms of each series of fungible phenomena are above such a high pointer and, for many phenomena, their series begins below the pointer, so a poor person finds that the possession of even a sample of such phenomena is beyond his means. Hence, a poor person possesses very little phenomena but what he has individually mean a lot to him, the marginal utility of each series of phenomena being given by the (high) position of his pointer or, for phenomena which comes in discrete units, the position of the lowest term in the series still above his pointer.

Wealthy people, on the other hand, have a pointer very low on their value scale. Even the most trivial phenomena can get at least a few terms of their series above such a low pointer. For most phenomena, such a person has more than he really knows what to do with. Hence, a wealthy person possesses a great deal of phenomena but what he has individually mean little to him. It is natural to ask whether a person is happier poor or rich. This is not the same thing as comparing the happiness of a poor person with a wealthy one because, since everyone has a unique value scale, such a question is meaningless. But the first question is legitimate, though some readers may feel the answer is obvious; they prefer to be rich. In the limit, however, the answer is indeterminate since, for a poor person, the value of individual phenomena approaches infinity as their quantity approaches zero, the multiplication of which is 0/0, unknown. Similarly for a wealthy person, the value of individual phenomena approaches zero as its quantity approaches infinity, the multiplication of which is again unknown. Indeed, complaints of unhappiness can be heard from all quarters of society, rich and poor.

What is needed is a function giving the number of definitions at each point on one's value scale so that one can integrate it from the position of one's pointer up to infinity and multiply the result by the utility denoted by the position of the pointer. This product will be called one's aggregate utility (happiness) and one can use standard calculus techniques to find its maximum value. Notice that this function is positive over its entire domain (one's value scale, which is itself always positive), and that it must converge. These two conditions imply that the needed function is a distribution. It is the distribution of stock on utility.

An immediate problem with this method is in choosing the unit for each phenomenon, because a small unit for a phenomenon causes the number of units in that series above the person's pointer to be disproportionally large, giving that kind of phenomenon extra weight. This problem is easily corrected, however, by observing that aggregate utility is maximized for a person only if it is maximized for each series of phenomena, so one can treat each series independently. In this case, the size of a unit is immaterial as a finite constant multiple on a function does not affect the position of that function's maximum.

The problem of defining a distribution which gives the position of each term in the series for a phenomenon can be broken into two parts: A function giving the position of the first term of the series, and one giving how much less each successive unit is worth relative to the last one, that is, a function giving diminishing utility. For the community, the Central Limit Theorem can be cited to define the distribution of first unit demand and the function giving people's value of later units can then be used to derive their demand distribution. For an individual, however, recourse cannot be made to the Central Limit Theorem, so it is impossible to say at exactly what level of wealth one will find the most happiness. Without a collective, one has no basis for assuming that a particular distribution is more accurate than another. Some will be called avaricious and some will be called lazy, but all must seek their station in society through trial and error. Philosophy is no guide, for it largely consists of someone tacitly assuming a particular distribution of stock on utility (usually his own as revealed through introspection) when, in fact, without a collective, one cannot prefer any particular distribution to another.

Section 10

10.1 The definition of the term "liquidity"

Some phenomena have a very low exchange value because of characteristics which make them hard to sell. One such characteristic is indivisibility, making the stime too great a purchase for many people or too drastic a sale to consider when only small gaps appear in the high end of one's value scale. Another such characteristic is one's eccentricity in considering a phenomenon to have use value. If one is unique among one's associates in giving it importance, one will have a difficult time selling it. Also, phenomena that only have value when associated with other complementary phenomena cannot be sold out of the set for a high price. Furthermore, many phenomena cannot be sold used because of people's fear that they are contaminated, worn out, or stolen. Phenomena like this that are difficult to sell have use value very much greater than exchange value, so their use value must drop a great deal before their exchange value becomes greater, necessitating a sale. Reflecting on this state of affairs, one who has lost interest in such phenomena and is eager to begin new ventures might wish for phenomena with a higher exchange value. The measure of exchange value relative to use value is liquidity.

With the advent of the division of labor, most phenomena are produced by people who have little use for them and want to exchange them for other phenomena, most of which have little liquidity. For this reason, they first exchange what they have for something more liquid, which the producers of those other phenomena need for the same reason. Thus, people value phenomena with a high exchange value relative to use value specifically for their liquidity. This further increases their exchange value until eventually a phenomenon is valued almost entirely for its exchangeability. This phenomenon is money. One keeps a proportion of one's wealth in money in anticipation of future expenditures. Some transactions take time to complete or set up, so one holds money in reserve for them so that one will be ready when the time comes. Other expenditures come up unexpectedly, so one must have a reserve of money to pay for them when they appear.

To qualify as money, a phenomenon must be capable of purchasing phenomena of less liquidity than itself. After people have selected one or two phenomena as money, the liquidity of other phenomena is still important for reasons that will be discussed in the next section. However, if liquidity measures only the ease of exchanging phenomena for money, then even very liquid phenomena are not money because they are not used to directly cancel debts. Thus, the terms "money" and "very liquid phenomena" are not synonymous. A phenomena can be very liquid if it is easy to sell (for money) but it can only be money if it is easy to use for purchasing all other types of phenomena, including those of low liquidity. Some bonds are very liquid because the central bank has promised to purchase all that are brought to it, but they do not function as money because they are not accepted as payment in private debts. They must be cashed first.

The value of money is a function of what one anticipates that it can buy in the future, this being a function of what it could buy in the past, adjusted for speculation on any changes that might affect its exchangeability. Each day, the value of money is determined by the demand for it by people who expect it to be exchangeable in the future. This supposition is based on its exchangeability in the past for, if people had not so easily accepted it as a medium of exchange yesterday, no one would expect them to do so tomorrow. Thus, each day the value of money is a proportion of its value the day before, regressing all the way back to the time when it was still valued primarily for personal use. The gradual rise in value is attributed to people's speculation that, judging from its past success, money would continue to spread throughout the world, replacing all other mediums of exchange. The natural rise in the value of money is a slow process which need concern no one but theoretical economists. However, the next chapter introduces institutions which greatly increase the practical importance of the daily speculation on the future value of money and the resulting price-change premiums added when a depreciation is expected.

That the value of money tends to increase as civilization advances may at first seem contrary to fact. If a person from a city visits a small, isolated community that still largely engages in direct exchange, he finds that only the wealthier people in that community use money, so the few monetary units that he brings with him command respect and can buy a larger quantity of phenomena than they would in the city. Thus, city people come to expect lower prices on locally producible items in their visits to the country which, though the savings do not offset the cost of the trip, are appreciated. Were civilization to advance, tearing down the political and transportation barriers to the less-advanced communities, country people would no longer consider money a rarity and would set their prices according to those charged in the city. People from the city who were used to finding low prices in the country would be disappointed and might conclude that the advance of civilization tends to devalue their money. Country people selling their wares in the city increase the stock there, however, lowering its price and counteracting the devaluation of money in the once-isolated community. This effect is less noticeable as it is spread out over a larger community while the smaller town goes through a more dramatic change.

Rather than saying that money has a different purchasing power in isolated communities, it is more appropriate to say that phenomena in one community are of a different kind than phenomena in another community isolated from the first, though they may appear identical. Thinking of them as being the same is a confusion caused by thinking of phenomena as objects rather than events. The events of perceiving spatially-separated phenomena are not the same, so they are different phenomena. For phenomena in isolated communities to be made identical with physically-similar phenomena in another community, the complementary phenomena "means of transport" and "payment of tariffs" must be added to them. Only when the addition of this complementary phenomena has become trivial are the communities assimilated. The same mistake is also at the root of many misconceptions regarding interest that presuppose temporally-separated phenomena to be identical because of their physical similarity.

All money that was originally valued for its usefulness and is physically limited in quantity, either because of people's inability to produce more of something which does not depreciate or their inability to prevent depreciation of something which is easily produced, is called commodity money. Let us say that the value of a unit of commodity money was V_0 when it was still valued primarily for its use value. After the jth day, the change in value is a proportion of the day before's value. Thus, $V_j - V_{j-1} = \varepsilon_j V_{j-1}$ with ε_j the change in value on the jth day relative to its value on the day before. If ε_j were the same every day, the value of commodity money would be growing exponentially with a common ratio of $1+\varepsilon$. ε_j is not the same every day, however, because each day has its own particular effect on the value of commodity money. In the long run, the ε_j's may be considered random as they are not directly related to each other nor are they uniquely a function of V_j. Furthermore, it is assumed that they are taken from an unspecified distribution, but one with a finite mean and a non-zero, finite average. Phenomena that change in this way every day conform to the characteristics of proportionate effect. Money certificates (see section 15.1) never had use value but are currently or were at one time associated with commodity money, and it is that time to which their value regresses.

There is no reason that the stock of money should change in proportion to the income of a community, that is, that its supply (supply is the first derivative of stock; the terms are often confused in economics textbooks) should be tied to the second derivative of a nation's wealth. It is one result of my theory that a loose monetary policy increases national income only by liquidating a nation's wealth, so national income cannot be considered the first derivative of national wealth. But even granting mainstream economists the assertion that the stimulus to national income from issuing money comes from somewhere other than a nation's wealth, it is difficult to see any connection between the second derivative of a nation's wealth and how much money should be issued. In fact, price adjustments allow any stock of money to work for any community as long as the stock of money is not subject to catastrophic changes. Changes in the stock of money redistribute a nation's wealth to those who get the new money but, if the changes are not too great, the money continues to function as a medium of exchange.

That there is no theoretical justification for tying the issue of money to national income is just as well, because neither the supply of money nor the income of a community can be measured empirically; no meaningful numbers can be associated with either parameter. It is well known that national income is a shaky statistic, but a nation's money supply is often discussed as though it were a value that their central bank adjusts at will. Central banks alone control the issue of money but, for relatively stable currencies, part of the stock is held in foreign countries (because their own currencies are unstable) and its flow across the border is unpredictable. For unstable currencies, the availability of more stable ones and the penalties for using them are unpredictable. Furthermore, central banks do not issue money *ad libitum* and then let commercial banks loan what they get. Rather, commercial banks make loans and then ask their central bank to cover them. Because of the central bank's commitment to prevent bank failures, it can rarely refuse such requests directly but can only chastise imprudent banks and incrementally raise their costs on future loans.

A central bank can peg the short-term interest rate with some degree of accuracy. However, the primary mission of a central bank is to support its government, so commercial banks often must jump through hoops when making loans to businesses rather than keeping their assets in government bonds (or home mortgages, which governments support because home-ownership makes people law-abiding). This imposes more telling costs on commercial banks than interest rates alone, but costs which are difficult both to measure and to control. Thus, as a central bank's regulatory bureaucracy over-corrects for past excesses or deficiencies, it often appears that the money supply is beyond their control.

For these reasons, every proposal to associate the supply of money with changes in national income has degenerated into arguments over empirical measurements. If only for rhetorical purposes, economists should advocate freezing the monetary base (or attaching it to a commodity money) and let price adjustments take their course. The monetary base is something that a central bank can control, while vague statistics on the money supply are not. Any argument for manipulating the money supply is weakened by the assumption that it can be controlled in a measurable way. In general, statistics on the money supply and on the measures of an economy that are affected by it are flexible enough to support virtually any issue and distribution of new money.

10.2 The value of complementary phenomena

It often happens that phenomena may conform to several definitions individually, but these definitions are conjoined into an alternative so that, if phenomena conform to all of them simultaneously, they complement each other in such a way that together they have more value than the sum of their independent conformances. In such cases, one endeavors to possess such phenomena in the proportions to each other that are required by one's theory, without having extras of any of those phenomena, for such singletons are of little value without their mates to complement them. The value of such singletons is a function of their liquidity and that is a function of how many different alternatives the definition to which a singleton conforms may be conjoined in and their importance. If a definition is never conjoined in any but one alternative and that by only a few people, one who finds oneself with a phenomenon conforming to that definition for which one cannot find the phenomena to complement it will have a difficult time selling it, for it is likely that one's colleagues have the same problem, perhaps having singletons themselves. Phenomena that conform to a definition that can be conjoined in a multitude of alternatives are more liquid, for, if the complementary phenomena required by an alternative cannot be found, it is probable that those required by another alternative can be found by someone and the singleton sold to him.

Whatever the relative liquidity of the phenomena conforming to the several definitions conjoined in an alternative, the value of any one of them while conjoined is the value of the set minus the individual values of all the others. This is because, when contemplating selling a member of a set, what is lost is the value of the set, though this loss can be partially redeemed by selling the remaining members. It may seem paradoxical that, where several phenomena with little liquidity contribute to a complementary set, they each have very close to the value of the set while conjoined. Thus, if one were to add up these values, one would arrive at a sum of the parts greater than the whole. This apparent contradiction is easily cleared up, however, when one remembers that the several members of a set cannot simultaneously have such high values for they are caused by breaking up the set and it can only be broken up once. Having assigned a high value to the member whose departure broke up a set and having allowed it to depart, the value of the set can no longer affect the value of the remaining members as it has already been lost. One can only speak of a sum of values if they can all be enjoyed. Contradictions arise when one allows mutually exclusive phenomena to be assigned values that can be added and subtracted from each other.

There is usually a large variance in the liquidity of the several phenomena conforming to the definitions in an alternative. If the more liquid ones are lost, leaving only those with low individual values because of their lack of liquidity, the owner will want to replace the lost members quickly, for he has lost the value of the set and it cannot be redeemed by selling the remaining members, for they have low individual values. Fortunately, the liquidity of the lost members makes them easy to replace, so the loss passes, according to the theory of marginal utility, to another part of the owner's wealth. If the less liquid, fixed member is lost, it is likely that, rather than replacing it, the owner will sell the remaining members. For this reason, it is common that one subtracts from the value of a set the individual values of the liquid members as one's costs, calling the difference one's reimbursement for owning the fixed or less liquid phenomenon.

The individual values of the members of a set can never amount to more than the value of the set, for if they did, one would employ them (or sell them) all individually and not consider them to form a set. This assertion (that individual values never amount to more than the value of a set) is always true, though one must not neglect the sentimental value owners often attach to a complete set. Where all of the members of a set are only moderately liquid, an owner facing bankruptcy must make the decision to sell the individual members while he still has time to make such sales, considering the low liquidity of the individual members. If he hesitates, he may later be forced to sell the complete set to a liquidator for less than the liquidator eventually realizes from selling the individual members. Either way, individual values never amount to more than the value of a set, though the value of a set may include sentimental values which do not appear in the accounts.

When all of the members of a set are very liquid, it may be that, were they all lost, they could all be replaced out of one's wealth for less than the value of the set. Thus, though the phenomena functioning as a set may satisfy an important want, it is only valued at the cost of its components. This is an easy application of the theory of marginal utility. It has, however, led some people to wonder, on observing that the value of phenomena is equivalent to their costs, how costs are determined. The answer is their product, though, according to the definition of liquid phenomena, they can have many products (combine into many sets), and thus it is their marginal product which determines their value. The value of each product is a function, though indirectly through their costs, of this marginal product. This confusion rarely occurs with less liquid phenomena for they have only one product and are valued directly for it.

Section 11

11.1 Indifference between a phenomenon and money

I assert that one is capable of determining which of any two phenomena or sets of phenomena conform to a definition at a higher place on one's value scale than the other. If one fails to determine which of the two is higher, it can only mean that they are equal. In other words, one's value scale is a total (linear) ordering of phenomena. For readers without a background in mathematics, this assertion may seem trivially true of all orderings. However, consider the concept of "subset", which is an ordering (it is transitive, reflexive, and antisymetric) but is not total; that is, it is not always true that, of two sets, one is a subset of the other.

This assertion is the first of three axioms which the reader is asked to accept. The plausibility of this axiom is derived mainly from analogy with the other dimensions (space and time), which are also totally ordered. A total ordering is included in the assertion of Absolute Geometry that every line has a coordinate system. This axiom is the principle motivation for Chapter One, which provides arguments for the autonomy of our will and value scale from God. Notice that if people were driven by spirits and could receive guidance through prayer, their value scales would not necessarily be totally ordered and most of economic theory would be reduced to second-guessing Divinity.

Because of this axiom, for every definition on one's value scale to which phenomena might conform, there stands beside it the number of units of money to which one is indifferent as to which one received. This supposition demands only that money be infinitely divisible, which it is for all practical purposes. As there are an infinity of distinct points on one's value scale, however, it cannot be expected that one is conscious of them all. In fact, one does not need to know exactly what one's point of indifference is to conduct many transactions. Most transactions are either definitely beyond one's reach or so mundane as to require no thought. Only for an individual on the margin, whose point of indifference is so close to the price that he must carefully weigh the alternatives before making a decision, does the point of indifference have an exact meaning. If one were to ask that individual in a less stressful situation, he would not know his point of indifference. In addition to not knowing, he would have a strong motive to lie. If he possessed a unit of the phenomenon in question which he valued only slightly, but thought the questioner valued highly, he might pretend to value it almost as highly in an effort to negotiate a sale in the narrow zone of indeterminacy thus created rather than the wider one which in fact exists. Also, knowing that someone is interested may make him suspect a larger market than he had previously imagined and prompt him to adjust his exchange value accordingly. Even watching him participate in the market only supplies information about the past, his valuations being changed by having bought or sold some of his phenomena in the market as well as by all the other changes wrought by the passage of time. Furthermore, the information supplied is very vague, indicating only that his valuation was greater than or less than a price but not even hinting at how much greater than or less than that price it was, for it cannot be known if the transaction came easily to him or if his point of indifference was very near the price asked.

As it is impossible to determine the quantity of money that a person holds indifferent to a phenomenon, one should not attempt to hypothesize each individual's point of indifference, but rather try to determine the distribution of points of indifference in the population. This does not make the data any more reliable, however, and one should not expect to realize the accuracy enjoyed by other sciences. One should be suspicious of any economist who too easily draws long, continuous lines on his graph paper, explaining away any doubts with the statement that they were curve fit, as though his critics were all ignorant of regression theory. One should rather ask him how he gets any points at all for his graphs, and whether their temporal separation is great enough to represent different situations. In general, no more can be known about a specific situation than is known about all situations, that is, no more than what can be deduced from universal axioms.

11.2 The interaction of stock and demand

The graph of the distribution of points of indifference, c(m), can be pictured as an aerial view of the people who value a phenomenon assembled along a line marked "money", where they are asked to stand by the number of monetary units that are equal to a unit of that phenomenon. If more than one person has the same valuation, they stand behind the corresponding number. The stock of that phenomenon naturally tends toward the high end, as anyone who possesses a unit of it who sees his neighbor to the right without one will sell it to him. Only use-value and expected exchange-value in other markets not represented on this graph are counted because, though one may value a phenomenon greatly in anticipation of exchanging it at a high price, if one fails to get that price, one has to lower one's asking price until it eventually equals the value of keeping that phenomenon for one's personal use. While money has very little use-value, it does have expected exchange-value in other markets not represented on this graph, and it is with this in mind that people withhold their money from this market if the price rises too high. The expected exchange-value of money is historically derived from its use-value as explained in section 10.1. If it were a function of today's prices, we would have a contradiction because we are now deriving today's prices from the demand distribution, c(m), which includes expected exchange-value.

If a phenomenon has a steeply-diminishing utility for most people (after acquiring one unit, the importance of the next is very low because one easily becomes sated), most people are only represented once and c(m) is very close to $c_0(m)$, the distribution of people's point of indifference for their first unit. If there is a gradually-diminishing utility among people, many come back again and again before they become sated, each time with a lower point of indifference, and consequently the low end of c(m) rises. See figure 8. $R = \int_0^\infty c(m)dm$ is the requirement for a phenomenon by a population. Because stock is limited, however, only those with the highest use-value of it relative to their value of money possess any of the phenomenon. The price is less than the point of indifference of the last person who possesses a unit of

Figure 8

The Demand Distribution, c(*m*)

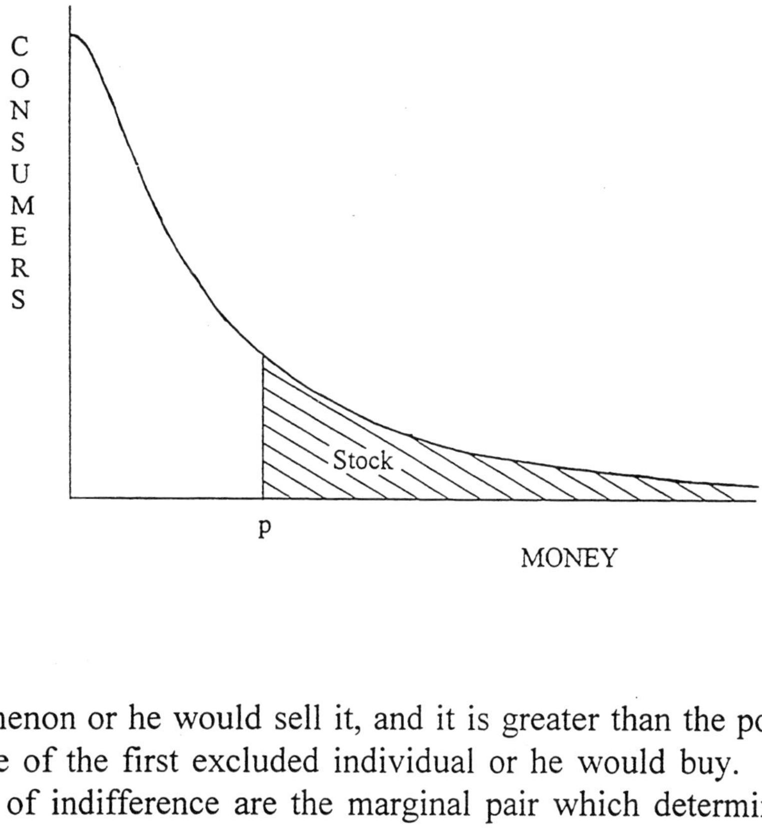

the phenomenon or he would sell it, and it is greater than the point of indifference of the first excluded individual or he would buy. These two points of indifference are the marginal pair which determine the upper and lower limit of price, between which is the zone of indeterminacy. The formula relating price and stock to the demand distribution, c(*m*), is $S(m) = \int_{m}^{\infty} c(t)dt$ with S(*m*) the stock, *m* the price, and c(*t*) (*t* is a dummy variable for the integration) the distribution of points of indifference between the use-value of a unit of a phenomenon and *t* units of money. S(*m*) is the survival function of the demand distribution; the function mapping stock onto price is the inverse survival function. Of course, the expression above does not have any meaning until it is proven that stock converges. It will be used informally, however, until section 11.3 when the convergence of stock is proven.

Both the people traditionally labeled "consumers" and those labeled "producers" appear in the demand distribution. The conceptual separation of consumers and producers is a great mistake of mainstream economics. They are all just people, each with a bit of the stock, and they are all prepared to sell if the price is above a certain point and buy if the price is below that point. The only thing that distinguishes people from one another is their point of indifference. This has little to do with who produced different bits of the stock, the event of production having occurred in the forgotten past. When economists draw one curve called "supply" and another called "demand", they are implying that the two are independent, for one cannot solve two simultaneous equations for two variables if the two equations are just versions of the same relation. Their dependence is well known at the macro level, but I assert that supply and demand are not independent at the micro level either. It is a mistake to inquire whether I support Say's assertion that "supply creates its own demand" or Keynes' assertion that "demand creates its own supply"; **Axiomatic Theory of Economics** is detached from that debate. I anticipate that the greatest block to the understanding of my theory will be people trying to interpret it in terms of supply and demand. I do not believe in supply and demand. I believe in the demand distribution, which is a mapping between price and stock. Supply has no place at all in **Axiomatic Theory of Economics**. My theory is not even divided into "micro" and "macro" sections. These terms were invented by mainstream economists when it became necessary to paste Keynes' theory over the top of Marshall's theory. They are clearly incompatible and their association in modern textbooks is entirely due to the bookbinder, not the economist.

By what criterion does mainstream economics distinguish people represented on a supply curve from those represented on the associated demand curve? This is a particularly pressing question for people dealing in narcotics because the penalties are so much greater for being on one curve than the other. But, if one visits a neighborhood where such trade takes place, any of the people one encounters would sell if the price were right and would buy if offered a bargain. There is really only one relation and it is called the demand distribution. Since there are two variables, price and stock, this (single) relation can provide a mapping from one variable to the other but cannot fix them. However, section 12.1 establishes the existence of a point toward which price and stock tend. Thereafter, it will be assumed that they are fixed at that point, called saturation.

The method of mainstream economics really has a third variable which is never mentioned and that is the time unit for supply and demand. It is well known that elasticity is a function of this time unit and, if this is true, one calculates a different price depending on whether one speaks of weekly or monthly supply and demand. This is an inconsistency since there can only be one price and it is not dependent on the caprice of an economist when he decides how often to conduct his surveys. This is a point which is glossed over in mainstream texts. A detailed discussion of the time unit chosen for supply and demand is never given and many texts neglect to mention the need for choosing one at all. Yet in their chapter on elasticity, every textbook lists time as a factor, sometimes as the most important factor.

Mainstream economists have two variables, price and quantity per unit of (some usually unspecified) time, and two equations, supply and demand. For this to work at all, the equations must be independent, which means that each individual must be either a buyer or a seller. The economist's decision to put people on one curve or the other cannot depend on the price that they would buy or sell at because both equations are defined for all prices. (Price is one of the independent variables.) So what is the economist's decision based on? Ask him repeatedly until he admits that there is really only one distribution. Also, press him to acknowledge that the demand distribution independently exists at each instant of time. Supply and demand curves are different depending on the time unit chosen. Mainstream economists provide no proof that their predicted prices are independent of their choice of time unit. For example, will thirteen predicted weekly quantities be the same as three predicted monthly quantities?

A large part of the problem with supply and demand is that it is used descriptively, but called predictive. It is easy to predict the past.

A large part of the problem with supply and demand is that it is used descriptively, but called predictive. It is easy to predict the past. Economists just observe the quantity produced one month and what it sold for and they put a little × over that spot. Then, by pure conjecture, they draw four tails on their × to fill their graph paper. Supply and demand has never been used predictively, not even to make bad predictions. × marks the spot is a purely descriptive technique. Since they are using the 20-20 vision of hindsight, they can do this for three months in a row and, to nobody's surprise, the sum of the quantities is the quarterly quantity. In the real world, price is constant for years at a time but, for most companies, their weekly and monthly sales figures swing wildly and unpredictably, sometimes by several fold from one month to the next. Mainstream economists have no explanation for this, which they should since their theory is called supply and demand and the horizontal axis of their graph is labeled weekly (or monthly) quantity. When I have been asked to help predict sales, I have told them that price is related to stock, not supply, and that they should stop watching their sales chart so ardently. At most companies, there is someone in accounting who feeds sales figures to the employees so that they can predict layoffs. They know that every dip in sales will send hundreds of them to the unemployment office, and that every rise will have their bosses clapping each other on the back and extolling their brilliant and farsighted management. They also know that nobody can predict sales. Supply never means anything in economics, though sometimes (for non-durable phenomena) it can pass for stock.

Working people have noticed that they are sometimes harried and sometimes bored. The cause of the one or the other has been the source of endless lunchroom speculation. The fact that their bosses also do not know has cost them prestige, especially the MBAs who tend to pepper their speech with talk of supply and demand curves, as though that meant something. I tell them that economists, including myself, have nothing to contribute to this conversation. Price is related to stock, not supply. Regarding the volatility of quantity sold per unit of time, it is best to regard it as a random variable. As such, it would have a Poisson distribution, which implies that its variance is equal to its expectation. If $q_1, q_2, \ldots, q_n$ are the quantities sold per unit of time of n independent phenomena and $p_1, p_2, \ldots, p_n$ are their associated prices, then total sales per unit of time, $\sum_{i=1}^{n} p_i q_i$, has the expectation $\sum_{i=1}^{n} p_i E(q_i)$. Since variance equals expectation for each independent random variable, the variance of total sales per unit of time is $\sum_{i=1}^{n} p_i^2 E(q_i)$. If the sample variance of 30 or more successive time units does not significantly exceed the population variance predicted above, then it is futile to invoke supply and demand to explain random fluctuations. Except in cases of unusually rapid growth or decline, changes in quantity sold per unit of time can almost always be explained as random fluctuations. (For people who are actually going to do this, make sure that the several phenomena are independent. If they complement each other or are substitutes for each other, the math is messier.)

250 years ago, when economics was first studied systematically, supply and demand made more sense. The only organized market was for agricultural products and, even there, the market did not operate continuously. Once a week, everybody came to town to attend church and to buy and sell food products. Without refrigeration, nothing lasted from one week to the next and so supply coincided with stock. Since the market was only held once a week, economists were not trying to be deceptive when they shortened "weekly quantity" to "quantity". Perfect competition, as the term is used by mainstream economists, was fairly descriptive. After the industrial revolution, supply and demand no longer worked. Industrial products are durable and it is their stock, not their supply, that is related to price. Unfortunately, economists continued to use their theory and to label the horizontal axis of their graphs "quantity" (but to mean weekly quantity) even though markets now operated continuously and the week was an arbitrary time unit. Late into the industrial revolution, tiny villages surrounded by family farms were still very common. As long as economists chose their examples carefully, it was not immediately apparent that they had gone astray. Today, it is apparent to everybody who gives the subject some thought. It is truly appalling that, as I write

these words, the 21st century approaches and supply and demand is still the undisputed theory of mainstream economics.

It is well known that mainstream economics is in trouble. Nobody in the hard sciences respects economists and even within their own ranks, a number of books and articles have appeared questioning why economics is not yet a science. There is considerable debate among economists about methodology, what it takes to qualify as a science, and what distinguishes economics from other fields. Implicit throughout is the understanding that mainstream economics does not work. Keynes' theory lost credibility in the 70's and, in the past two decades, no new theorist has stepped forward to replace Keynes. The several modern schools (Monetarist, Rational Expectationist, New Classical and New Keynesian) are all just minor modifications of Keynesianism, though marked by the cautious tone that economists learned to use after the debacle of stagflation. In **The Crisis of Vision in Modern Economic Thought**, 1995, Heilbroner and Milberg write[10]:

> *The prestige that attaches to economics may far outweigh its achievements, but there is certainly cause for disquiet when the branch of social inquiry most closely associated with political and social policy lacks a firm foundation, in the sense of a core set of beliefs that commands more or less universal assent from both within and outside the profession.*

That Heilbroner, a popularist for the profession, would write a book like **Crisis of Vision** is an indication of just how much trouble economics is in. This is made especially clear when the book concludes without offering a new theoretical approach, but only (*ibid*, page 119) "a recognition of the necessity for a widening degree and deepening penetration of public guidance into the workings of capitalism itself." Like the self-help manuals peddled at grocery stores, **Crisis of Vision** switches cause and effect between one's failures and one's lack of confidence (vision); if only economists will act like they know what they are doing when they intervene in the economy and unite in their call for more government intervention, they might get it. This psychological response to one's failures is similar to the effect so often seen in old people, whose growing senility is outpaced only by their arrogance. In **The Death of Economics**, 1997, Ormerod writes[11]:

> *Whatever solution is offered, whatever prognostication is made, one quality above all is lacking from the various utterances. Namely, that of humility. For the record of economists in understanding and forecasting the economy at the macro-level is not especially impressive. Indeed, uncharitable writers might be inclined to describe it as appalling.*

While hard sciences like physics are still waiting for another Einstein to unify field theory, the practical success of physicists assures them that such a man will only extend modern physics, just as Einstein extended Newtonian Mechanics. But he will not refute their work; no new observations will defy the Correspondence Principle. This is not the case in economics, where every discussion of a new paradigm fills economists with dread. They are nearly always wrong when using the current paradigm and have no assurance that all of their work may not be discarded. Yet methodology continues to be debated because economists know that, without drastic changes, their status as scientists will eventually have to be abandoned.

Economists in this debate are roughly divided into three camps: Those who believe that the problems are caused by inappropriate use of statistics, those who believe that the problems are caused by inappropriate use of deduction, and those who believe that, as long as government grant money is flowing in, there is no problem. From a practical point of view, the third group is right. Questions of methodology and the meaning of the word "science" are idle because, at least in the United States, "scientific" is all that is paid for by the National Science Foundation and "unscientific" is all that is not. By that definition, **Axiomatic Theory of Economics** is unscientific. From a more idealistic point of view, the problems are caused by both inappropriate use of statistics and of deduction. I have already discussed statistics in Chapter Three. As for deduction, I assert that economics is no different from sciences where deduction has proven useful. It is just that, until now, the axioms of economics were faulty and vague. Much of what is considered deductive in economics is actually a charade, with economists paying lip service to vague, non-mathematical axioms from which nothing is actually deduced, at least not in the mathematical sense of the word. Witness the great many f(x)'s floating through economics texts which are never actually specified so that mathematicians can deal with them.

To qualify as a science, economics must be axiomatic. But one must address price and stock; supply and demand does not work. Also, to deduce mathematical expressions from axioms, the axioms must be of a mathematical nature and they must specify actual functions from which equations can be derived. Economics is an applied science and one does not choose one's axioms only for the richness of the theory which can be deduced from them (otherwise my book would have more theorems), but so that phenomena can conform to them and people can tell when they do. One tests phenomena for conformance to one's axioms and then one assumes that phenomena conform to the implications of those axioms. One does not arbitrarily assume that one's axioms are true and then test phenomena for conformance to the implications of those axioms. What would be the point of the latter method? If the predicted phenomena were observable, one would just observe them and forget about theory.

Fortunately, however, there is nothing fundamental about economics that prevents it from being made into a science just as physics was made into a science by Newton and mathematics by Euclid. Only a few hundred years ago it was common to segregate assertions into relations of ideas, such as the Pythagorean Theorem, and matters of fact. An example of the latter was given by David Hume in 1748[12]:

> *When I see, for instance, a billiard ball moving in a straight line toward another,... may I not conceive that a hundred different events might as well follow from that cause?... All these suppositions are consistent and conceivable. Why then should we give preference to one which is no more consistent or conceivable than the rest?*

However, the term "consistent" is only meaningful in reference to a specific set of axioms, in this case those of Newtonian Mechanics. Only one of the hundreds of different events mentioned in the above quotation conserve both momentum and energy, so, in reference to these two axioms, all the other events are inconsistent. The purpose of having a theory at all is that one does not have to apply experience to every event but need only apply it once when deciding on one's axioms, in this case that momentum and energy are both conserved. After that, application of the theory is algorithmic. The only reason that the assertion of how billiard balls will react to a collision was considered a matter of fact and the Pythagorean Theorem was considered a relation of ideas is because Euclid's five postulates were older then Newton's laws of physics at the time. As it turned out, both sets of axioms were shown to be only one of several possible systems, each intuitive in its own way and each internally consistent. Economics is younger now than physics was in 1748 and no one of Newton's stature has been attracted to it. But nothing prevents finding a set of axioms for economics just as Newton found a set of axioms for physics. This is what I propose to do in **Axiomatic Theory of Economics**.

11.3 Analysis of the demand distribution

There is an upper bound to one's value of any stock of a phenomenon which will be denoted M. This includes one's need for saving phenomena for future use. Total utility is the marginal utility of a phenomenon when the unit is defined as the entire quantity possessed. It increases from zero up to its maximum point, M, as one's stock increases. Hence, total utility is a cumulative distribution function and marginal utility is the associated probability density function (after normalization), denoted $U(s)$ and $u(s)$, respectively. Since the utility of a given stock is measured by the quantity of money which stands beside it on one's value scale, $U(s)$ is a mapping from the stock of a phenomenon one possesses to the money one associates with that stock. $u(s)$ is its first derivative. $u(s)$ must be negative monotonic because utility diminishes as one adds units to one's stock. The integral of $u(s)$ must also converge, that is, $\int_{0}^{\infty} \mathrm{u}(s)ds < \infty$. This is because marginal utility is the probability density function of a cumulative distribution. Nothing else is known about u(s) and this is the first parameter (and the only function) used to distinguish phenomena from one another. Its characteristics must be regarded as an axiom. Later in this section two more parameters (both from $\Re^{+}$) will be introduced which will be sufficient to completely describe every phenomenon. A graph of U(s) and u(s) is given in figure 9.

As will be shown shortly, we are only concerned with the ratio $\frac{\mathrm{u}(0)}{\mathrm{u}(r)}$ for non-negative integers, r. This ratio is invariant under a rescaling of the vertical axis, so u(s) can be normalized by setting the upper bound on the distribution function, M, to unity. This makes u(s) a true probability density function as the total area under it is unity.

It should be noted here that the requirement that u(s) be negative monotonic does not imply that firms must be small, which is clearly not true because there are many large and successful corporations. Economists have used the term "marginal (or diminishing) utility" to denote both the first derivative of one's total utility for some phenomenon and the assertion that firms receive less and less return on their investments as they grow bigger. Capital, like all phenomena, has diminishing utility because one quickly becomes sated on it. However, like most things that one temporarily sates oneself on, one is ready for more the next day and the day after that. Thus, while a firm cannot immediately make use of all the capital it might consider buying, it can start with a small capital project and use the profits from that to train the managers and laborers that will make an expansion feasible. In this way, firms can become global in scale without ever contradicting the assertion that u(s) is negative monotonic for capital. The large corporation embarking on another great expansion may have started out as a small mom-and-pop outfit, but it is not that little company anymore and it has a (very) different utility function now. Since **Axiomatic Theory of Economics** is about stock, not supply, the relative sizes of the firms supplying a phenomenon is of no concern.

Figure 9

The Relation Between Total and Marginal Utility

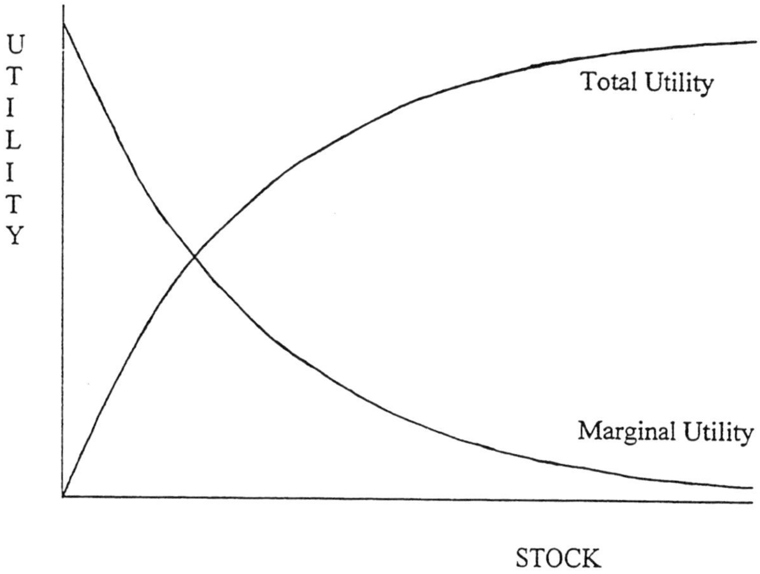

I assert that the distribution of people's points of indifference for their first unit of a phenomenon relative to money, $c_0(m)$, is lognormal; that is, the natural logarithm of the number of people who are indifferent at a particular price, *m*, is cumulatively (normally) distributed. The cumulative distribution is applicable to a variable that is subject to a process of change such that, at each step, a random quantity is added to the accumulated value of that variable. By the Central Limit Theorem, the distribution of the sum of a large number of independent, identically-distributed random variables (from an unspecified distribution with a finite mean and a non-zero, finite variance) is approximately normal. $c_0(m)$, however, does not accumulate, rather it is analogous to the growth of the value of money through history: It conforms to the characteristics of proportionate effect. After the *j*'th day of a person's life, the change in the number of monetary units to which he is indifferent, relative to the first unit of a phenomenon, is a proportion of his indifference point the day before. That anthropometric variables (height, size of organs, tolerance to drugs, etc.) conform to the characteristics of proportionate effect is well established in the literature.

Theorem 1 (Law of Proportionate Effect): Phenomena which conform to the characteristics of proportionate effect are lognormally distributed.

Proof:

$m_j - m_{j-1} = \varepsilon_j m_{j-1}$	The difference between each step and the last one is the last one multiplied by a random quantity.
$\dfrac{m_j - m_{j-1}}{m_{j-1}} = \varepsilon_j$	Divide through by m_{j-1} to get ε_j, the change in m relative to its previous value, m_{j-1}.
$\displaystyle\sum_{j=1}^{n} \frac{m_j - m_{j-1}}{m_{j-1}} = \sum_{j=1}^{n} \varepsilon_j$	Find the sum of all ε_j from the initiation of the process to its termination after n steps.
$\displaystyle\int_{m_0}^{m_n} \frac{dm}{m} = \sum_{j=1}^{n} \varepsilon_j$	If each step is small, $m_j - m_{j-1}$ can be approximated by dm.
$\ln\lvert m_n\rvert - \ln\lvert m_0\rvert = \displaystyle\sum_{j=1}^{n} \varepsilon_j$	Integrate from m_0 to m_n.
$\ln\lvert m_n\rvert = \ln\lvert m_0\rvert + \varepsilon_1 + \ldots + \varepsilon_n$	Solve for $\ln(m_n)$.

As can be seen from the last step, the natural logarithm of one's indifference point after the n'th day is a constant (the logarithm of its initial quantity) with a large number of random and identically-distributed quantities accumulated onto it. Hence, after having lived through n days and having seen their point of indifference change by a small proportion each day, consumers of their first unit are normally distributed with regard to the variable $\ln(m)$ and, hence, are lognormally distributed with regard to the variable m. ■

The absolute value operation may be dropped, since we are only interested in positive prices.

That first-unit demand conforms to the characteristics of proportionate effect must be regarded as an axiom. A plausibility argument is provided here. Let $m_j = \phi(m_{j-1})$ with m_j the number of monetary units to which one is indifferent relative to the first unit of a phenomenon on the j'th day of that person's life. We want to show that $\phi(m_{j-1}) = (1+\varepsilon_j)m_{j-1}$. Consider a man who wants to take out a loan at interest. He must think he will have more money in the future than he does now. (More money holdings, not necessarily more wealth.) If he does, the value of individual monetary units will tend to decrease over time relative to other phenomena; that is, ϕ is a positive function when averaged over all phenomena. To determine how much interest he is willing to pay, the man must specify this average ϕ. For him to calculate the interest owed per unit of time as a percentage of the principle is equivalent to specifying $\phi(m_{j-1}) = (1+\varepsilon)m_{j-1}$ with $\varepsilon > 0$ fixed. Fixing ε is a special case of ε_j being a random variable. Here, the probability density function is unity at ε and zero elsewhere. Thus, the axiom that first-unit demand conforms to the characteristics of proportionate effect is a generalization of calculating interest as a percentage of the amount owed. In fact, this is how people have calculated interest throughout recorded history, although economics having always been a soft science, they never asked for proof. Perhaps the value of money decays harmonically over time or in another way besides exponentially? This question is addressed in Appendix A, but for now let us proceed to investigate the consequences of people's points of indifference for their first unit of each phenomenon being lognormally distributed. I believe that this axiom is on solid intuitive ground and will not be criticized. Even if it is, it is unlikely that critics will succeed in convincing the banking industry to calculate interest with a different formula, so the weight of tradition will continue to support my choice of the lognormal distribution for first-unit demand.

Before continuing, let us explicitly state our three axioms:

1) One's value scale is totally (linearly) ordered:

- **i)** Transitive; $p \leq q$ and $q \leq r$ imply $p \leq r$
- **ii)** Reflexive; $p \leq p$
- **iii)** Antisymetric; $p \leq q$ and $q \leq p$ imply $p = q$
- **iv)** Total; $p \leq q$ or $q \leq p$

2) Marginal (diminishing) utility, u(s), is such that:

- **i)** It is independent of first-unit demand.
- **ii)** It is negative monotonic; that is, $u'(s) < 0$.
- **iii)** The integral of u(s) from zero to infinity is finite.

3) First-unit demand conforms to proportionate effect:

- **i)** Value changes each day by a proportion (called $1+\varepsilon_j$, with j denoting the day), of the previous day's value.
- **ii)** In the long run, the ε_j's may be considered random as they are not directly related to each other nor are they uniquely a function of value.
- **iii)** The ε_j's are taken from an unspecified distribution with a finite mean and a non-zero, finite variance.

ln(m) is linearly transformed by $\frac{\ln(m)-\mu}{\sigma}$. The location parameter, μ (mean), quantifies the importance of a phenomenon relative to money and the scale parameter, σ (standard deviation), quantifies the difficulty of substituting other phenomena for the one in question. Easily-substituted phenomena have very little probability in the tail of their demand distribution; only the eccentric purchase a phenomenon at a high price when there are cheaper substitutes available. As substitution becomes more difficult, people must purchase the phenomenon even at high prices, and their distribution is less skewed. Both μ and σ must be positive. With u(s), μ and σ describe all phenomena. Thus, every phenomenon is associated with a point in u(s),μ,σ space where u(s) is a negative-monotonic probability density function on $\Re^+$ and μ and σ are both from $\Re^+$. For the purpose of economics, nothing else distinguishes one phenomenon from another.

The equation for the distribution of first-unit demand is $c_0(m) = \frac{e^{-\frac{1}{2}\left(\frac{\ln(m)-\mu}{\sigma}\right)^2}}{\sigma m}$. This is the equation of the lognormal distribution, $e^{-\frac{\ln^2(m)}{2}}$, multiplied by the derivative of the linear transformation which is substituted for $\ln(m)$. It need not be divided by its total area, $\sqrt{2\pi}$, since it will not be used as a probability density function.

The number of people with a point of indifference at a particular price, m, for their first unit is $c_0(m)$. Whoever's point of indifference for his first unit is $\frac{u(0)}{u(1)}$ times greater than that price values his second unit equivalent to price m. Whoever's point of indifference for his first unit is $\frac{u(0)}{u(2)}$ times greater than that price values his third unit equivalent to price m, and so on. To find $c(m)$, all the people with a point of indifference at m are summed up, whether it is their first purchase or a later purchase. Recall the analogy of the demand distribution being an aerial view of the people who value a phenomenon assembled along a line marked "money", where they are asked to stand by the number of monetary units that are equal to a unit of that phenomenon. Now consider a person who wishes to possess more than one unit of the phenomenon; each of his agents appears behind a different point on the money line. If he himself appears in the column assembled behind m monetary units, the first person he sends to get another unit is directed to the column behind $\frac{m u(1)}{u(0)}$ monetary units. His next agent is in the column behind $\frac{m u(2)}{u(0)}$ monetary units, and so on. Hence, we have the following formula for the demand distribution which, unfortunately, is impossible to integrate in closed form, even with $u(s)$ fixed.

$$c(m) = \sum_{r=0}^{\infty} c_0(x) \qquad \text{with } x = \frac{mu(0)}{u(r)}$$

$c_0(m)$ can be thought of as the 0'th partial sum of $c(m)$ and, in general, $c_n(m)$ denotes the n'th partial sum of $c(m)$. Thus,

$$c_n(m) = \sum_{r=0}^{n} c_0(x) \qquad \text{with } x = \frac{mu(0)}{u(r)}$$

Until the convergence of this summation is proven, $c(m)$ will be used informally.

As one suspects, most of the demand for an easily substituted phenomenon is concentrated at or just above the price because people do not get excessively attached to phenomena that can be replaced in other markets. This makes for quite a peak around the mode of $c_0(m)$, which lower points on the demand distribution may or may not include in their summations, as they hit only intermittent points. Thus, for easily substituted phenomena, $c(m)$ is oscillatory, gaining amplitude and losing frequency until just past the mode of $c_0(m)$, after which the tail out to infinity is negligible. The oscillations are primarily a function of substitutability because that determines how sharp a peak $c_0(x)$ has, but it is also a function of how quickly one becomes sated on a phenomenon, as that determines how intermittently $c_0(x)$ is polled by the summation. The demand distributions of phenomena for which it is difficult to find substitutions look more like conventional, unimodal, skewed distributions, except that they increase abruptly from zero. It almost appears that they have a positive value at the edge of their domain. Fortunately (for economists), the oscillations are most pronounced at the low end among people who are excluded from making a purchase, so changes in the parameters describing phenomena usually do not cause any unexpected changes in price or stock. In general, one can visualize the high end of the demand distribution (where the stock is) like the unimodal, skewed distributions one is familiar with from probability. The effect of substitutability on the shape of the demand distribution is illustrated in figure 10.

Let us begin our analysis with a theorem that will be referred to in several of the later proofs. Recall that u(s) is in the function space of negative-monotonic probability-density functions on $\Re^+$.

Theorem 2: There exists an N such that, for all n > N, $u(n) < \frac{1}{n}$.

Proof: Assume that for every N there exists an n > N such that $u(n) \geq \frac{1}{n}$. Then there also exists an m > 2N such that $u(m) \geq \frac{1}{m}$. Since u(r) is negative monotonic,

$$\sum_{r=N}^{\infty} u(r) > \sum_{r=N}^{m} u(r) > \sum_{r=N}^{m} \frac{1}{m} > \frac{1}{2}.$$

For the summation of u(r) to converge, for every ε > 0 there must exist an N such that $\sum_{r=N}^{\infty} u(r) < \varepsilon$. But if $\varepsilon < \frac{1}{2}$, no such N could exist, which contradicts what is known about u(s). ■

Figure 10

The Demand Distribution, c(*m*), with Large Sigma

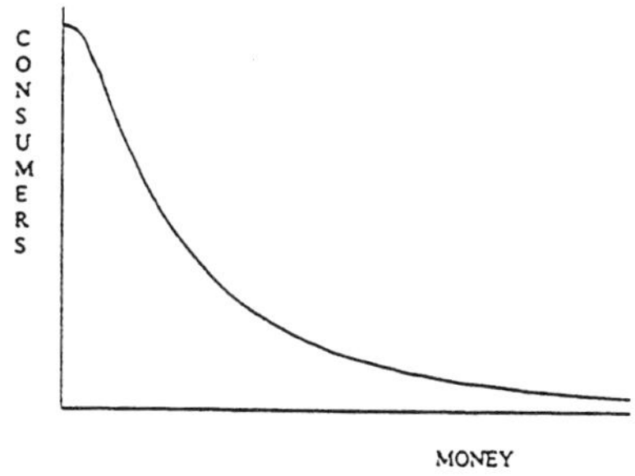

The Demand Distribution, c(*m*), with Medium Sigma

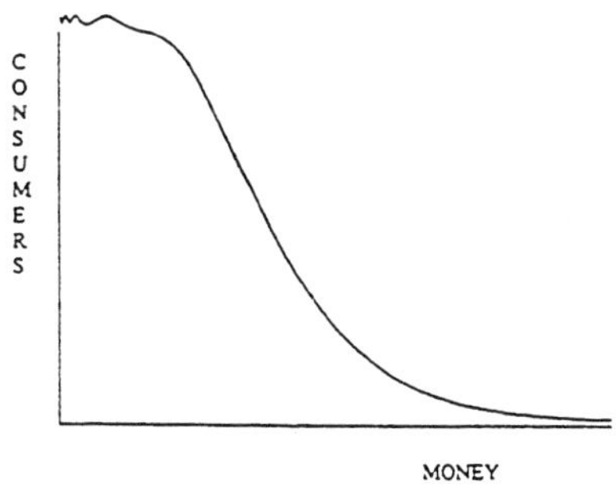

The Demand Distribution, c(*m*), with Small Sigma

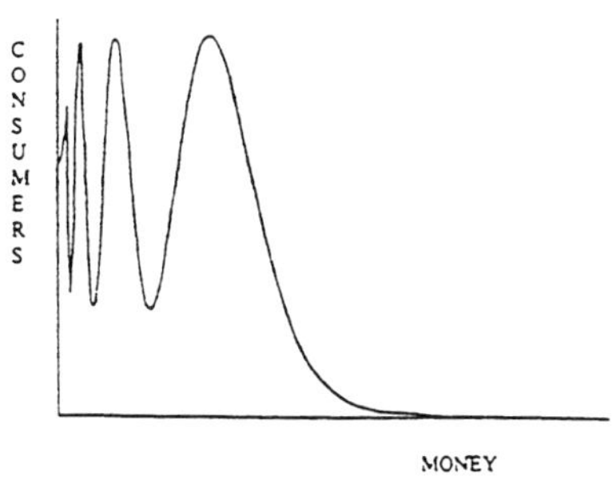

From here on, the smallest integer that satisfies this condition will be denoted N* By the Well-Ordering Property, N* exists and is unique.

Theorem 3: The demand distribution is u.c. convergent over $\Re^+$, that is, it converges uniformly over any compact subset of $\Re^+$.

Proof: Fix any p > 0 and let N' be the smallest integer such that $pu(0)N' > e^{\mu-\sigma^2}$ with $e^{\mu-\sigma^2}$ the mode of $c_0(x)$, after which it is strictly decreasing. For all N > max(N*, N') with N* defined in Theorem 2 and $m > p$,

$$\sum_{r=N+1}^{\infty} c_0\left(\frac{mu(0)}{u(r)}\right) < \sum_{r=N+1}^{\infty} c_0\left(\frac{pu(0)}{u(r)}\right)$$

$$< \sum_{r=N+1}^{\infty} c_0(pu(0)r)$$

$$< \frac{1}{pu(0)}\int_N^{\infty} e^{-\frac{y^2}{2}}\,dy$$

It is well known that the tail of the normal distribution can be made arbitrarily small with large enough N. ■

Practically, this means that the BASIC program in Appendix B (more will be said about it in section 12.1) can predetermine N so that the calculation of the demand distribution will have an arbitrary degree of accuracy regardless of what the price at saturation turns out to be, as long as it is greater than a minimum price, say, one monetary unit.

Theorem 4: $\lim_{m\to 0^+} c_0(m) = 0$

Proof: $c_0(m) > 0$ for all $m > 0$. Thus, by the Squeezing Theorem, if $c_0(m)$ is less than some function for all $m > 0$ and that function is continuous and equals zero at zero, then $\lim_{m\to 0^+} c_0(m) = 0$. Consider h$m$ with h a finite constant. Since hm vanishes at zero, it is sufficient to show that $hm > \dfrac{e^{-\frac{1}{2}\left(\frac{\ln(m)-\mu}{\sigma}\right)^2}}{\sigma m}$ for all $m > 0$. By making the substitution $y = \ln(m)$, this is equivalent to $y^2 + (4\sigma^2 - 2\mu)y + 2\sigma^2\ln(\sigma h) + \mu^2 > 0$ for all real y. By the Quadratic Theorem, this is true for $\dfrac{e^{2(\sigma^2-\mu)}}{\sigma} < h < \infty$. Thus, the demand distribution is equal to zero at zero. ■

Alternate proof: Make the substitution $y = \ln(m)$ so

$$\lim_{m\to 0^+} c_0(m) = \lim_{y\to -\infty} \frac{e^{-\frac{1}{2}\left(\frac{y-\mu}{\sigma}\right)^2}}{\sigma e^y}$$

$$= \frac{1}{\sigma}\lim_{y\to -\infty} e^{-\frac{(y-\mu)^2 + 2\sigma^2 y}{2\sigma^2}}$$

$$= \frac{1}{\sigma}\lim_{y\to -\infty} e^{-\frac{(y-\mu+\sigma^2)^2 - \sigma^2(\sigma^2 - 2\mu)}{2\sigma^2}}$$

$$= 0 \quad ■$$

The former was chosen as the main proof because the Squeezing Theorem and the Quadratic Theorem can be visualized and are (hopefully) more intuitive than a purely algebraic proof. It is an easy corollary that the limit of $c(m)$ at zero is zero.

Theorem 5: The uniform convergence of the demand distribution can be extended to zero.

Proof: To extend uniform convergence to zero, it is sufficient that $\lim_{x\to 0^+} c_0(x) = 0$. If it equaled $L > 0$, then for any finite N, $n \geq N$

implies that, for any $\varepsilon > 0$, $\sum_{r=n+1}^{n+j} L = jL > \varepsilon$ for some integer j. The existence of an integer, j, such that $jL > \varepsilon$ for positive L and ε is implied by the Archimedean Principle.

In other words, if one sets p to be the smallest unit in which it is practical to measure prices, then by Theorem 3, the approximation of $c(m)$ by an N term summation is within ε of its correct value for all $m > p$ and, by Theorem 4, so is $c(0)$ because $0 + \ldots + 0$ N times is zero no matter what N is. ■

Theorem 6: The integral in the equation for stock can be brought inside the summation.

Proof: In order to calculate stock, one must first bring the integral operation inside the summation. In the context of Lebesque integration, this would be an easy application of the Monotone Convergence Theorem. However, in the context of Riemann integration, the problem is more difficult. If the summation were finite, this step would not require any justification since it is well known that the integral of a finite sum is the sum of the integrals. The equation for stock actually involves the limit of a finite sum, $\lim_{A\to\infty} \int_m^A \lim_{n\to\infty} \sum_{r=0}^{n} c_0(x)dt$ with $x = \frac{tu(0)}{u(r)}$.

Thus, the problem is reduced to bringing the integral operation inside a limit, since, having done that, it is a trivial matter to bring it inside a finite summation.

The proof is divided into two parts. First, if stock has a compact support (A is finite), uniform convergence of the demand distribution is sufficient. Second, the tail from A to ∞ can be made arbitrarily small so that one may indeed assume that stock has a compact support.

i) If there is an N such that $n \geq N$ implies that $\int_m^A c(t)dt - \int_m^A c_n(t)dt < \varepsilon$ for each fixed $\varepsilon > 0$, the difference between the integral of the limit and the integral of the n'th partial sum can be made arbitrarily small by making N large enough, so they are equal at infinity. By Theorem 3, the demand distribution converges uniformly, so there is an N such that $n \geq N$ implies $0 < \sum_{r=n+1}^{\infty} c_0(x) = c(t) - c_n(t) < \frac{\varepsilon}{A-m}$ with $x = \frac{tu(0)}{u(r)}$ for all $t > m$ and $m < A < \infty$. Thus, for all $n \geq N$,

$$\int_m^A c(t)dt - \int_m^A c_n(t)dt = \int_m^A c(t)-c_n(t)dt$$

$$< \int_m^A \frac{\varepsilon}{A-m}dt$$

$$= \varepsilon$$

Therefore,

$$S(m) = \lim_{A,n\to\infty} \int_m^A \sum_{r=0}^{n} c_0(x)dt \qquad \text{with } x = \frac{tu(0)}{u(r)}$$

$$= \lim_{A,n\to\infty} \sum_{r=0}^{n} \int_m^A c_0(x)dt \quad \square$$

ii) Notice that $A < \infty$ is essential for bringing the integral operation inside the summation. So it only remains to be shown that the integral from A to infinity can be made arbitrarily small with a large enough A, still finite.

$$c_0(x) = \frac{e^{-\frac{1}{2}\left(\frac{\ln(x)-\mu}{\sigma}\right)^2}}{\sigma x} \qquad \text{with } x = \frac{t\mathrm{u}(0)}{\mathrm{u}(r)}$$

$$= \frac{e^{-\frac{\ln^2(x)-2\mu\ln(x)+\mu^2}{2\sigma^2}}}{\sigma x}$$

$$= \frac{e^{-\frac{\mu^2}{2\sigma^2}}\, e^{-\frac{\ln^2(x)}{2\sigma^2}}}{\sigma x x^{-\frac{\mu}{\sigma^2}}}$$

$$= \frac{1}{\sigma e^{\frac{\mu^2}{2\sigma^2}}\, x x^{\left(\frac{\ln(x)}{2\sigma^2}-\frac{\mu}{\sigma^2}\right)}}$$

$$\leq \frac{1}{x^2 \sigma e^{\frac{\mu^2}{2\sigma^2}}} \qquad \text{if } t \geq e^{2(\mu+\sigma^2)}$$

$$= \frac{\left(\frac{\mathrm{u}(r)}{\mathrm{u}(0)}\right)^2}{t^2 \sigma e^{\frac{\mu^2}{2\sigma^2}}}$$

$$\leq \frac{\frac{u(r)}{u(0)}}{t^2 \sigma e^{\frac{\mu^2}{2\sigma^2}}}$$

$$= \frac{Cu(r)}{t^2} \qquad \text{with } C = \frac{1}{\sigma e^{\frac{\mu^2}{2\sigma^2}} u(0)}$$

Thus, for $A \geq e^{2(\mu+\sigma^2)}$,

$$\int_A^\infty c(m)dm = \lim_{B\to\infty} \int_A^B \sum_{r=0}^{\infty} c_0(x)dt \qquad \text{with } x = \frac{tu(0)}{u(r)}$$

$$\leq \lim_{B\to\infty} C \int_A^B t^{-2} \sum_{r=0}^{\infty} u(r)dt$$

$$= \lim_{B\to\infty} CD \int_A^B t^{-2} dt \qquad \text{with } D = \sum_{r=0}^{\infty} u(r)dt$$

$$= CD \lim_{B\to\infty} \left(\frac{1}{A} - \frac{1}{B}\right)$$

$$= \frac{CD}{A}$$

C and D are both finite constants, so $\frac{CD}{A}$ can be made arbitrarily small by increasing A. □

Thus, the integral in the equation for stock can be brought inside the summation. ■

Theorem 7: Stock is finite.

Proof: By Theorem 6, the integral in stock can be brought inside the summation. Make the substitutions $y = \frac{\ln\left(\frac{t u(0)}{u(r)}\right) - \mu}{\sigma}$ and $dy = \frac{dt}{\sigma t}$ so

$$S(m) = \sum_{r=0}^{\infty} \frac{u(r)}{u(0)} \int_{z}^{\infty} e^{-\frac{y^2}{2}} dy \qquad \text{with } z = \frac{\ln\left(\frac{m u(0)}{u(r)}\right) - \mu}{\sigma}$$

The integral inside the summation is the standard normal distribution, which is tabulated as $\alpha(z) = 1 - \Phi(z)$ in the back of any statistics text, though multiplied by the constant $\frac{1}{\sqrt{2\pi}}$ so that the total area under the integrand is unity, a step which is omitted here since the integrand is not being used as a probability density function. However, since this integral never exceeds $\sqrt{2\pi}$ and u(0) is a constant, we have the following inequality: $S(m) < \frac{\sqrt{2\pi}}{u(0)} \sum_{r=0}^{\infty} u(r)$. Since marginal utility converges, by the Integral Test for convergence, stock converges also. With an appropriate choice of units, it can be scaled up to any finite number. ■

Stock actually converges considerably faster than marginal utility because the integral not only never exceeds $\sqrt{2\pi}$, but goes to zero quite rapidly. Thus, the rate of convergence of stock is largely independent of the distribution chosen to represent marginal utility. The convergence of stock is an important result; all of the later proofs depend on it.

Finding marginal utility, u(*s*), seems a simple matter as most dealers post price lists for varying quantities bought and one could divide the price by the quantity to get the unit price of a phenomenon at that quantity. Doing this at the several different quantities that the price per unit changes gives one enough points to fit the curve, u(*s*), to it. This does not determine the rate that price per unit decays against quantity possessed, however, but only its lower bound. The amount of discount that one gives for quantity purchases is uniquely determined by one's need to sell one's inventory quickly. Diminishing utility (marginal utility - the word "diminishing" is used to emphasize that it is negative monotonic) has no direct influence on the rate of discount except, were it set at too gradual a rate, it would be ineffective and nobody would buy ahead. Therefore, if one observes a price list posted and people buying in quantity, the only thing that one can learn is that people's diminishing utility is less than the price list or they would not be buying in quantity. Hence, the equation for u(*s*) must be considered very obscure. Even more obscure is how the distribution of people's point of indifference concerning the first unit of a phenomenon, $c_0(m)$, is linearly transformed. This is because it is difficult to determine the importance of phenomena relative to money and to quantify the substitutability of a phenomenon.

Far more serious errors are created, however, by plotting points on the price/stock graph whose observations were separated in time enough that the demand distribution, c(*m*), cannot be considered constant between them. In fact, that stock changed at all may be an indication that the parameters of c(*m*) changed between successive observations and that it was this change and not external events which prompted the change in stock. Later considerations show that stock tends toward a unique point (saturation) and thus, if the parameters of the demand distribution remain constant for any length of time, one can expect stock to settle at that point. Thus, variances in stock indicate changes in the underlying parameters, so one cannot measure prices at different stocks and expect c(*m*) to remain constant throughout.

Because the parameters of the demand distribution, c(m), are difficult to measure, numerical estimates of how changes in price or stock affect each other cannot be made with any accuracy beyond the usual platitudes. In general, no more can be known about a specific situation than is known about all situations, that is, no more than what can be deduced from the three axioms. In economics, it is a mistake to extend the implications of one's axioms with statistical inferences so that different predictions are made in different empirical situations. There is no justification for the assumption that price and stock are inversely proportional (have a slope of -1) or even linear over a significant portion of their graph, that is, that c(m) is a rectangular distribution. This mistaken assumption is particularly evident in many theories concerning the stock of money and its purchasing power, which naively presume this linearity. Index numbers notwithstanding, purchasing power is probably the most obscure of all quantities in economics.

Section 12

12.1 The definition of the term "saturation"

Aggregate utility is defined as price multiplied by stock. This is because money is the measure of utility and everyone who possesses a unit of stock values it only as highly as its replacement cost, for that is all that one risks. Stock and price, however, are inversely related, so increasing one or the other does not necessarily increase aggregate utility. Aggregate utility being the common goal of people dealing in a phenomenon, they are interested in maximizing it. As stock increases, aggregate utility also increases up to saturation, where any further increases in stock reduce aggregate utility by driving the price down. That part of the demand distribution to the right of saturation (the high end), where increases in stock increase aggregate utility, is unsaturated and that part to the left (the low end) is saturated. At a constant stock, there is a zone of indeterminacy between the marginal pair within which the price may fluctuate. Such fluctuations appear to be of a saturated market whether the stock has reached saturation or not. Most markets are large enough, however, that the zone of indeterminacy is too narrow to be of practical concern.

Because the actions appropriate in an unsaturated market (increasing stock) are not those appropriate in a saturated market (decreasing stock), it is important to determine the point of saturation. Aggregate utility, $m\mathrm{S}(m)$, is at a relative maxima where its first derivative, $\mathrm{S}(m) - m\mathrm{c}(m)$, equals zero. Thus, saturation is a price and stock such that $\mathrm{S}(m) = m\mathrm{c}(m)$. Graphically, $\mathrm{S}(m)$ is represented by the area between the horizontal axis and the graph of the demand distribution from m to ∞. $m\mathrm{c}(m)$ is represented by the area of the rectangle formed by the two axes and horizontal and vertical lines extending from the point $m,\mathrm{c}(m)$. See Appendix B.

Theorem 8: For all $m > \mathrm{p}$ with p an arbitrary positive constant, the first and second derivatives of $\mathrm{S}(m) - m\mathrm{c}(m)$ are

$$\sum_{r=0}^{\infty} c_0(x)\left(\frac{\ln(x)-\mu}{\sigma^2} - 1\right) \qquad \text{with } x = \frac{mu(0)}{u(r)}$$

and

$$\frac{1}{m}\sum_{r=0}^{\infty} c_0(x)\left(1 - \left(\frac{\ln(x)-\mu}{\sigma^2}\right)^2 + \frac{1}{\sigma^2}\right) \qquad \text{with } x = \frac{mu(0)}{u(r)}$$

Proof: Uniform convergence is sufficient for integrating inside a summation but not for differentiating inside a summation. This is because uniform convergence assures that the n'th partial sum, $f_n(m)$, is within an ε-band of $f(m)$ but it does not exclude the possibility of oscillations within that ε-band. In both intervals shown in figure 11, $|f_n(m) - f(m) < \varepsilon|$ for all m but only in the first is the derivative of $f_n(m)$ close to that of $f(m)$.

It is sufficient to show that for every $\varepsilon > 0$ there exists an N such that

$$\left|\sum_{r=N+1}^{\infty} \frac{df_0(r,m)}{dm}\right| < \varepsilon \qquad \text{with } f(m) = \sum_{r=0}^{\infty} f_0(r,m)$$

for all m in an interval with N possibly dependent on that interval. Thus, if the summation of the derivatives is uniformly convergent in an interval, $f'(m) = \sum_{r=0}^{\infty} \frac{df_0(r,m)}{dm}$ for m in that interval. Here we need the absolute value operation. It was ignored while proving that stock is finite because $c_0(x) > 0$ for all x.

Figure 11

Differentiating Inside a Summation

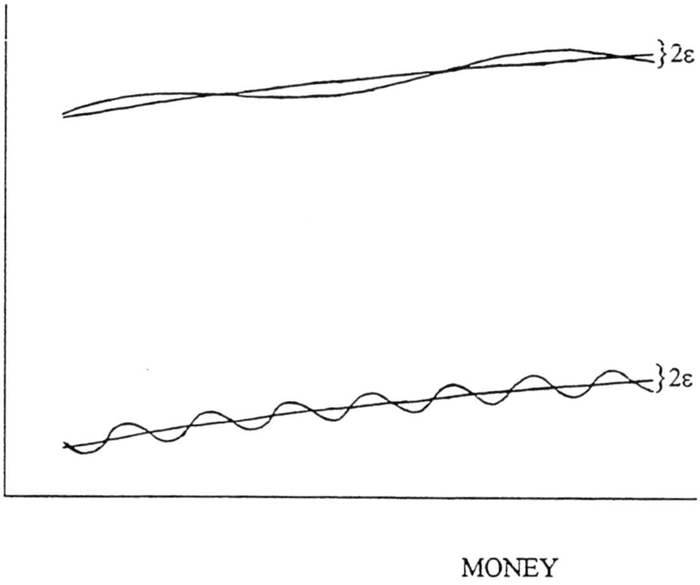

i) For the first derivative of $S(m) - mc(m)$,

$$\left|\frac{\ln(x)-\mu}{\sigma^2} - 1\right| = \frac{\ln(x)-\mu}{\sigma^2} - 1 \qquad \text{for } x > e^{\mu+\sigma^2}$$

$$< \frac{\ln(x)}{\sigma^2}$$

$$< \frac{x}{\sigma^2}$$

Let N' be the smallest integer such that $\mathrm{pu}(0)\mathrm{N}' > e^{\mu+\sigma^2}$ and N* be defined in Theorem 2. Notice that $e^{\mu+\sigma^2} > e^{\mu}$ with e^{μ} the mode of $x\mathrm{c}_0(x)$. For all N > max(N*, N') and $m >$ p,

$$\left| \sum_{r=N+1}^{\infty} \mathrm{c}_0(x)\left(\frac{\ln(x)-\mu}{\sigma^2} - 1\right) \right| < \sum_{r=N+1}^{\infty} \frac{x\mathrm{c}_0(x)}{\sigma^2} \qquad \text{with } x = \frac{m\mathrm{u}(0)}{\mathrm{u}(\mathrm{r})}$$

$$< \sum_{r=N+1}^{\infty} \frac{\mathrm{pu}(0)}{\sigma^2\mathrm{u}(\mathrm{r})}\mathrm{c}_0\left(\frac{\mathrm{pu}(0)}{\mathrm{u}(\mathrm{r})}\right)$$

$$< \sum_{r=N+1}^{\infty} \frac{\mathrm{pu}(0)\mathrm{r}}{\sigma^2}\mathrm{c}_0(\mathrm{pu}(0)\mathrm{r})$$

$$< \frac{1}{\sigma^2}\int_{N}^{\infty} y e^{-\frac{y^2}{2}} dy$$

$$= -\frac{1}{\sigma^2}\left[e^{-\frac{y^2}{2}} \right|_{N}^{\infty}$$

$$= \frac{e^{-\frac{N^2}{2}}}{\sigma^2}$$

which can be made arbitrarily small with large enough N. □

ii) For the second derivative of S(m) - mc(m),

$$\left|1 - \left(\frac{\ln(x)-\mu}{\sigma^2}\right)^2 + \frac{1}{\sigma^2}\right| = \left(\frac{\ln(x)-\mu}{\sigma^2}\right)^2 - 1 - \frac{1}{\sigma^2} \qquad \text{for } x > e^{\mu+\sigma^2\sqrt{1+\sigma^{-2}}}$$

$$< \frac{\ln^2(x)}{\sigma^4}$$

$$< \frac{x}{\sigma^4} \qquad \text{for } x > 1$$

Let N' be the smallest integer such that pu(0)N' > max($e^{\mu+\sigma^2\sqrt{1+\sigma^{-2}}}$, 1) and N* be defined in Theorem 2. Notice that pu(0)N' > max($e^{\mu+\sigma^2\sqrt{1+\sigma^{-2}}}$, 1) > e^{μ} with e^{μ} the mode of $xc_0(x)$. For all N > max(N', N*) and m > p,

$$\left|\frac{1}{m}\sum_{r=N+1}^{\infty} c_0(x)\left(1 - \left(\frac{\ln(x)-\mu}{\sigma^2}\right)^2 + \frac{1}{\sigma^2}\right)\right| < \sum_{r=N+1}^{\infty} \frac{xc_0(x)}{\sigma^4 m} \quad \text{with } x = \frac{m\text{u}(0)}{\text{u(r)}}$$

$$< \sum_{r=N+1}^{\infty} \frac{\text{pu}(0)}{\sigma^4\text{pu(r)}} c_0\left(\frac{\text{pu}(0)}{\text{u(r)}}\right)$$

$$< \sum_{r=N+1}^{\infty} \frac{pu(0)r}{\sigma^4 p} c_0(pu(0)r)$$

$$< \frac{1}{\sigma^4 p} \int_N^{\infty} y e^{-\frac{y^2}{2}} dy$$

$$= -\frac{1}{\sigma^4 p} \left[e^{-\frac{y^2}{2}} \right|_N^{\infty}$$

$$= \frac{e^{-\frac{N^2}{2}}}{\sigma^4 p}$$

which can be made arbitrarily small with large enough N. □

This shows that, when calculating the first and second derivatives of S(*m*) - *m*c(*m*), the infinite summations may be treated as finite summations, in which case the sum of the derivatives is the derivative of the sum. ■

Another derivative will now be established which is not needed until section 13.1. Its proof is included here because it is similar to the preceding proofs.

Theorem 9: For all $0 < \mu < b$ with b an arbitrary positive constant, the first derivative of S(μ,*m*) - *m*c(μ,*m*) with respect to μ is

$$-m\sum_{r=0}^{\infty} c_0(x)\left(\frac{\ln(x)-\mu}{\sigma^2} - 1\right) \qquad \text{with } x = \frac{mu(0)}{u(r)}$$

Proof: Differentiating S(μ,*m*) - *m*c(μ,*m*) with respect to μ is complicated by the integral in stock. By Theorem 6, this can be brought inside the summation. Having done that, to bring the derivative inside the integral requires an application of Leibnitz' Rule, that is, it requires that

$$\frac{dc_0(\mu,x)}{d\mu} = c_0(\mu,x)\frac{\ln(x)-\mu}{\sigma^2} \qquad \text{with } x = \frac{mu(0)}{u(r)}$$

(t is a dummy variable for the integration) be continuous. It is, indeed, continuous. To extend Leibnitz' Rule to an improper integral, however, it is also required that for every $\varepsilon > 0$ there exists an A such that $\left|\int_A^{\infty}\frac{dc_0(\mu,x)}{d\mu}dt\right| < \varepsilon$ for all $\mu < b$. That is, as the upper limit increases, the integral converges uniformly with respect to μ.

$$\left|\int_A^{\infty}\frac{dc_0(\mu,x)}{d\mu}dt\right| = \int_A^{\infty}\frac{dc_0(\mu,x)}{d\mu}dt \qquad \text{for } A > e^b$$

$$< \int_A^{\infty}\frac{dc_0(b,x)}{d\mu}dt$$

$$= \frac{u(r)}{\sigma u(0)} \int_z^{\infty} y e^{-\frac{y^2}{2}} dy \qquad \text{with } z = \frac{\ln\left(\frac{Au(0)}{u(r)}\right) - b}{\sigma}$$

$$= -\frac{u(r)}{\sigma u(0)} \left[e^{-\frac{y^2}{2}} \right|_z^{\infty}$$

$$= \frac{u(r)}{u(0)} \frac{e^{-\frac{z^2}{2}}}{\sigma}$$

z is a positive monotonic function of A so $e^{-\frac{z^2}{2}}$ can be made arbitrarily small with large enough A.

The remainder of the proof need only establish uniform convergence of the summation of the derivatives in an interval, which was all that was required in the proof of Theorem 8.

$$\left| \frac{\ln(x)-\mu}{\sigma^2} - 1 \right| = \frac{\ln(x)-\mu}{\sigma^2} - 1 \qquad \text{for } x > e^{b+\sigma^2}$$

$$< \frac{\ln(x)}{\sigma^2}$$

$$< \frac{x}{\sigma^2}$$

Let N' be the smallest integer such that $m\mathrm{u}(0)N' > e^{b+\sigma^2}$ and N* be defined in Theorem 2. Notice that $e^{b+\sigma^2} > e^{\mu}$ with e^{μ} the mode of $x c_0(x)$. For all N > max(N*, N') and μ > b,

$$\left| -m \sum_{r=N+1}^{\infty} c_0(\mu,x)\left(\frac{\ln(x)-\mu}{\sigma^2} - 1\right) \right| < \sum_{r=N+1}^{\infty} \frac{m x c_0(\mu,x)}{\sigma^2} \quad \text{with } x = \frac{m\mathrm{u}(0)}{\mathrm{u}(r)}$$

$$< \sum_{r=N+1}^{\infty} \frac{m^2 \mathrm{u}(0)}{\sigma^2 \mathrm{u}(r)} c_0\left(b\,,\ \frac{m\mathrm{u}(0)}{\mathrm{u}(r)}\right)$$

$$< \sum_{r=N+1}^{\infty} \frac{m^2 \mathrm{u}(0) r}{\sigma^2} c_0(b, m\mathrm{u}(0)r)$$

$$< \frac{m}{\sigma^2} \int_{N}^{\infty} y e^{-\frac{y^2}{2}} dy$$

$$= -\frac{m}{\sigma^2}\left[e^{-\frac{y^2}{2}} \right|_{N}^{\infty}$$

$$= \frac{me^{-\frac{N^2}{2}}}{\sigma^2}$$

which can be made arbitrarily small with large enough N. ■

Now, the existence of a saturation point for every phenomena will be proven.

Theorem 10 (existence): The absolute maximum of aggregate utility is at a finite critical point.

Proof: By Theorem 4, the limit of $c_0(m)$, and hence of $c(m)$, at zero is zero. Thus, stock is finite even if it is free, and aggregate utility goes to zero as price approaches zero. Since aggregate utility is always positive, it is sufficient to show that it also goes to zero as price approaches infinity to prove the existence of a relative maxima. By Theorem 6, the integral in stock can be brought inside the summation.

One then makes the substitutions $y = \frac{\ln\left(\frac{tu(0)}{u(r)}\right) - \mu}{\sigma}$ and $dy = \frac{dt}{\sigma t}$ so

$$0 < mS(m) = m \sum_{r=0}^{\infty} \frac{u(r)}{u(0)} \int_{z}^{\infty} e^{-\frac{y^2}{2}} dy \qquad \text{with } z = \frac{\ln\left(\frac{mu(0)}{u(r)}\right) - \mu}{\sigma}$$

$$< m \sum_{r=0}^{\infty} \frac{u(r)}{u(0)} \int_{\frac{\ln(m)-\mu}{\sigma}}^{\infty} e^{-\frac{y^2}{2}} dy$$

$$= Am \int_{\frac{\ln(m)-\mu}{\sigma}}^{\infty} e^{-\frac{y^2}{2}} dy \qquad \text{with } A = \sum_{r=0}^{\infty} \frac{u(r)}{u(0)}$$

$$\leq Am \int_{\frac{\ln(m)-\mu}{\sigma}}^{\infty} y e^{-\frac{y^2}{2}} dy \qquad \text{if } m \geq e^{\mu+\sigma}$$

$$= -Am \left[e^{-\frac{y^2}{2}} \right|_{\frac{\ln(m)-\mu}{\sigma}}^{\infty}$$

$$= Ame^{-\frac{1}{2}\left(\frac{\ln(m)-\mu}{\sigma}\right)^2}$$

$$= Ame^{-\frac{\ln^2(m)-2\mu\ln(m)+\mu^2}{2\sigma^2}}$$

$$= \frac{Bme^{-\frac{\ln^2(m)}{2\sigma^2}}}{m^{-\frac{\mu}{\sigma^2}}} \qquad \text{with } B = \frac{A}{e^{\frac{\mu^2}{2\sigma^2}}}$$

$$= \frac{\mathrm{B}m}{m^{\frac{\ln(m)-2\mu}{2\sigma^2}}}$$

$$\leq \frac{\mathrm{B}}{m} \qquad \text{if } m \geq e^{2(\mu+2\sigma^2)}$$

B is a constant, so $\lim_{m\to\infty} \mathrm{B}/m = 0$ and, by the Squeezing Theorem, $\lim_{m\to\infty} m\mathrm{S}(m) = 0$. Thus, there exists a finite price where aggregate utility is at a maximum. ■

This only proves the existence of a relative maxima and identifies it with the absolute maximum. There may be more than one relative maxima, in which case the largest of them is the absolute maximum. Because aggregate utility is always positive and it approaches zero at both ends of its domain, $(0,\infty)$, there is either a single relative maxima, or relative maximas and minimas alternate with the largest and smallest being relative maximas. The second derivative of aggregate utility is

$$\sum_{r=0}^{\infty} \mathrm{c}_0(x)\left(\frac{\ln(x)-\mu}{\sigma^2} - 1\right) \qquad \text{with } x = \frac{m\mathrm{u}(0)}{\mathrm{u}(r)}$$

It is positive at relative minimas and negative at relative maximas. Therefore, if there is more than one relative maxima, there are two disjoint intervals in $(0,\infty)$ where the second derivative is negative and they are separated by an interval where the second derivative is positive.

The proof that two such disjoint intervals do not exist could be divided into three parts:

1) The second derivative is positive for $m > e^{\mu+\sigma^2}$.

2) The second derivative is positive monotonic for

$$e^{\mu-\sigma^2\sqrt{1+\sigma^{-2}}} < m \leq e^{\mu+\sigma^2}.$$

3) The second derivative is negative for $m \le e^{\mu-\sigma^2\sqrt{1+\sigma^{-2}}}$.

$e^{\mu+\sigma^2}$ is the zero of the second derivative of $c_0(m)$ and $e^{\mu+\sigma^2\sqrt{1+\sigma^{-2}}}$ is a zero of the third derivative of $c_0(m)$. Proof of the first part is easy:

Theorem 11: $e^{\mu+\sigma^2}$ is an upper bound for any saturation price.

Proof: We wish to show where the second derivative is strictly positive. $c_0(m) > 0$ for all x, so we only have to examine $\frac{\ln(x)-\mu}{\sigma^2} - 1$. This is positive for all $x > e^{\mu+\sigma^2}$. Thus, since, $x \ge m$ for all r (because marginal utility is negative monotonic), $e^{\mu+\sigma^2}$ is an upper bound for any saturation price. ■

It is not difficult to prove uniqueness when it is assumed that, at a given time, nobody needs more than one of anything, that is, when one uses $c_0(m)$ rather than $c(m)$. This is done in the introduction. In general, however, the third part of the proposed proof fails for phenomena that are very easy to substitute for, that is, $\sigma \ll 1$. This is because, for small σ, $c_0(m)$ has a very sharp peak and, for r = 1, 2, ..., there are m_1, m_2, ... such that $\frac{m_r u(0)}{u(r)}$ hits the peak. So, for small σ, $c(m)$ is mostly very close to zero except for a sharp peak at $e^{\mu-\sigma^2}$ (the mode of $c_0(m)$) and several peaks less than it at m_1, m_2, ... Saturation is near a peak because, away from the peaks, $c(m) \approx 0$, so $mc(m) \approx 0$ and cannot equal $S(m)$. Thus, if the peaks are sharp enough, there is a price near each peak such that $S(m) = mc(m)$ and every other one is a relative maxima of aggregate utility. Where there are several relative maximas, however, it appears that there is never more than one greater than $e^{\mu-\sigma^2\sqrt{1+\sigma^{-2}}}$, which indicates that only the third part of the proposed uniqueness proof fails.

Appendix B gives the source code of a BASIC program which prints out a graph of the demand distribution and does a linear search for relative maximas from zero to $e^{\mu+\sigma^2}$. By Theorem 11, $e^{\mu+\sigma^2}$ is an upper bound for any saturation price. Readers may use this program to look for a counter-example to the assertion that there is at most one saturation price greater than $e^{\mu-\sigma^2\sqrt{1+\sigma^{-2}}}$. An informal justification is also included in the appendix. Unhappily, I have failed to prove this assertion, though I have not found a counter-example to it either, and continue to believe that it is true for all combinations of μ, σ, and $u(s)$. I leave the proof or refutation of this assertion as an exercise for the reader.

Despite the lack of a general uniqueness proof, I will continue to use the word "the" when speaking of the saturation point. There are several reasons for this:

1) Multiple relative maximas only occur for phenomena that are very easy to substitute for and, realistically, there are not many phenomena described by such small values of σ;

2) When there is more than one saturation point, it appears that all but one are at such low prices (less than $e^{\mu-\sigma^2\sqrt{1+\sigma^{-2}}}$) that it is unlikely that phenomena ever sell for so little; and

3) When there is more than one saturation price, aggregate utility is very near zero between them and it is unlikely that prices jump from one relative maxima to another.

This latter point is important because a principle result of this chapter (and of this book) is the Law of Price Adjustment, which addresses how prices change in response to changes in the importance of phenomena relative to money. It assumes that, as parameters gradually change, the price of a phenomenon follows the relative maxima that it is on. The assertion that prices do not jump from one relative maxima to another is empirical and, hence, contradicts my belief that it is a mistake to extend the implications of one's axioms with statistical inferences so that different predictions are made in different empirical situations. Readers are welcome to investigate the possibility of aggregate utility having more than one relative maxima in the vicinity of an observed price. The BASIC program in Appendix B should prove useful to such an investigation.

Most of the real analysis in **Axiomatic Theory of Economics** stems from the infinite summation, $c(m)$. All of the main results, including the Law of Price Adjustment (which the reader will be relieved to learn is the only remaining proof), are proven in the introduction for the simplified case of $c_0(x)$ rather than $c(m)$; that is, the second axiom is replaced with the assertion that people never need more than one of anything at a time. This assumption is neither accurate nor necessary, as all of the results of my theory can be (and are) proven in their full generality. However, some economists do not have the mathematical background necessary to read my book, so, for expository purposes, simplified proofs are provided in the introduction. Also, before my theory becomes accepted, it will receive cursory reviews, perhaps at the end of courses on mainstream economics. In this case, if a professor is sympathetic to my theory, he may wish to prove some of its assertions, but he will not have time to prove them in their full generality. As long as he mentions that the complete proofs do exist, his students can get the essence of my theory from the simplified proofs. The important thing for them to understand is that this theory is deduced from axioms.

There are three parameters which affect the position of the saturation point:

1) $u(s)$, marginal (diminishing) utility;

2) μ, the importance of the phenomenon relative to money; and

3) σ, the difficulty of substituting other phenomena.

Appendix B gives the source code of a BASIC program which prints out a graph of the demand distribution. The saturation point is marked by a rectangle, which should remind readers of how it is defined. Readers may wish to vary each parameter independently so that they can see its effect on saturation and the shape of the distribution.

Newton's method for finding a zero of a function is derived from the first two terms of its Taylor series approximation around a nearby point. One sets this to zero and solves for m. Then one repeats the process with the approximation around this new point which is (hopefully) closer to the zero of the function. To get more accuracy, one might use the first three terms of the Taylor series, but that could not be solved for m. To circumvent this, factor out an $(m - m_0)$ and solve for that m, approximating the other m with Newton's method. Hence, $m \approx m_0 - \dfrac{f(m_0)}{f'(m_0) - \dfrac{f(m_0)f''(m_0)}{2f'(m_0)}}$

For this method to work, the initial estimate, m_0, must be near the zero of f(m). By Theorem 12 (see section 13.1), one can set $\mu = 0$ and then multiply the estimated solution by e^{μ}. Arbitrarily fixing u(s) makes the solution of f(m) = 0 a function of σ only, that is, $f(\sigma,p(\sigma)) = 0$. $p(\sigma)$ cannot be expressed in terms of standard transcendentals but it has derivatives of all orders, starting with $-\dfrac{f_\sigma(\sigma,m)}{f_m(\sigma,m)}$, which indicates that one can approximate it with a Taylor series. Inspection reveals that, for large σ, $p(\sigma)$ increases at about the same rate as e^{σ^2}, however, which rules out any type of polynomial approximation.

To approximate $p(\sigma)$, I applied the Gram-Schmidt process to the sequence e^1, e^σ, e^{σ^2},... with the inner-product defined $f\|g = \int_0^\infty e^{-e^\sigma} f(\sigma)g(\sigma)d\sigma$. The first few terms of the resulting sequence are:

$$e_0(\sigma) = 2.13499857557$$

$$e_1(\sigma) = -4.86365405934 + 2.90042726954\, e^\sigma$$

$$e_2(\sigma) = 2.26806196853 - 2.24029715815\, e^\sigma + 0.883903147493\, e^{\sigma^2}$$

$$e_3(\sigma) = -1.7161607963E\text{-}12 + 1.69281159704E\text{-}12\, e^\sigma - 6.68243652657E\text{-}13\, e^{\sigma^2} + 1.68653454738E\text{-}22\, e^{\sigma^3}$$

p(σ) is approximated by $\sum_{n=0}^{2} p\|e_n\, e_n(\sigma)$. The inner-product is evaluated using Romberg integration.

Estimates of p(σ) are made using $c_0(m)$ and with u(*s*) fixed at $e^{\ln\left(\frac{1}{4}\right)\sigma}$, $e^{\ln\left(\frac{1}{2}\right)\sigma}$, and $e^{\ln\left(\frac{3}{4}\right)\sigma}$. The u(*s*) function input by the user at run-time is quantified by the geometric average of $\frac{u(i+1)}{u(i)}$ for i = 1, 2, 3,... and then a first-degree spline is used between the nodes 0, 1/4, 1/2, and 3/4.

Having estimated the solution to f(*m*) = 0, the only remaining obstacle is evaluating f ′(*m*) and f ″(*m*).

$$F(m) = mS(m)$$
$$f(m) = S(m) - mc(m)$$

$$f'(m) = \sum_{r=0}^{\infty} c_0(x)\left(\frac{\ln(x)-\mu}{\sigma^2} - 1\right) \qquad \text{with } x = \frac{mu(0)}{u(r)}$$

$$f''(m) = \frac{1}{m}\sum_{r=0}^{\infty} c_0(x)\left(1 - \left(\frac{\ln(x)-\mu}{\sigma^2}\right)^2 + \frac{1}{\sigma^2}\right) \qquad \text{with } x = \frac{mu(0)}{u(r)}$$

Justification for these derivatives is given in Theorem 8.

The most difficult part of this computation is estimating stock since the normal distribution function, Φ(z), cannot be calculated in closed form. There are many methods available for estimating the tail area, α, because of its use in statistics. For z < 1.28, the BASIC program uses the convergent series $e^{\frac{z^2}{2}}\int_0^z e^{-\frac{t^2}{2}}dt = z + \frac{z^3}{3} + \frac{z^5}{3\cdot 5} + \ldots$ which it solves for the integral from 0 to z, which is then subtracted from $\frac{\sqrt{2\pi}}{2}$ to get $\alpha(z)$. For $z \geq 1.28$, the BASIC program uses the continued fraction $e^{\frac{z^2}{2}} \int_0^z e^{-\frac{t^2}{2}} dt = \frac{1}{z+} \frac{1}{z+} \frac{2}{z+} \frac{3}{z+} \cdots$ which it solves for $\alpha(z)$. Given the required accuracy of seven decimal digits (the accuracy of single-precision BASIC), there is a faster method available which uses a polynomial approximation. However, it requires typing in 24 coefficients, which would be tedious for the reader. Also, it cannot achieve more accuracy in other languages, while the method used here is as accurate as the language it is coded in permits. This allows the user to change floating point packages or to translate the program into other languages without having to change any of the algorithms.

One might consider doing the integral recursively so $\Phi(z_r) = \Phi(z_{r-1}) + \int_{z_{r-1}}^{z_r} e^{-\frac{y^2}{2}} dy$ which would considerably shorten each interval of integration and allow one to use a standard integration method at each iteration. However, if marginal utility is very steep (it diminishes quickly), only a few iterations are necessary before the integral becomes negligible and, if marginal utility is very gradual, a great many iterations are necessary. Thus, the length of each interval and, hence, the number of function evaluations at each iteration, is dependent on u(s). Since u(s) is input at run-time, the BASIC program calculates each $\Phi(z_r)$ independently.

12.2 The effects of collusions meant to fix stock

People who deal in a certain phenomenon have a common interest at setting its stock at saturation. This serves to maximize the aggregate utility. If there is too little of a phenomenon, people hasten to make more of it. If there is too much of a phenomenon, people abandon its production in favor of more lucrative trades. This is a long-running process, but it can still be accomplished in the short run by an avaricious speculator increasing stock or a ruthless monopolist decreasing stock, respectively. The existence of speculation and monopolies is not an aberration. An economic theory that requires special artifices to explain what it considers aberrant behavior is no theory at all.

If stock exceeds saturation, the excess dealers are excluded from the market. The easiest people to exclude are foreigners, for a union can find an ally in the central bank, which also has an interest in preventing nationals from buying from foreigners. Unions and monopolists go a step further than the central bank, however, by preventing foreigners from immigrating into their geographical area. The next easiest class of people to exclude is foreigners who have only recently immigrated, because racial arguments against buying from foreign nationals are easily extended to buying from immigrants. This method presents a solid line of demarcation between outsiders and members of an alliance that is easily discernible even to people who do not understand the economics. The majority of immigrants are poor and, as they are excluded by the various alliances of labor and other phenomena upon arriving, they usually have a difficult time breaking out of their poverty. Thus, untraveled people tend to believe theories of racial superiority because the only people of foreign descent that they see are poverty-stricken. This reinforces discriminatory efforts so unions and monopolists usually meet with little opposition from excluded people. Economists who preface their theories with "Assuming perfect competition..." or “Assuming a continuum of traders...” are speaking of something that does not exist and, if it were defined more clearly, would probably be contradictory. It is not much of a theory which can be frustrated by something so mundane as people's efforts to set the stock of each phenomenon at saturation. The assumption of “perfect” competition (lots of tiny firms selling a homogenous product) is one of the five major mistakes of mainstream economics though, remarkably, it is not the worst of them. The others will be listed at the end of the book.

People who deal in a certain phenomenon are never interested only in increasing its price as some economists have asserted. They are interested in maximizing aggregate utility and, by definition, that means setting the stock at saturation. Politicians do not have as accurate an estimate of saturation as entrepreneurs, however, and tend to increase prices without regard to aggregate utility. When wholesale exclusions take place, such as the closing of the borders, consumers are often left with inefficient suppliers who happen to reside within the border. Failure to attain saturation is a symptom of monopolies set up along political lines rather than economic ones. At the other extreme, governments have sometimes emphasized the virtue of everybody making the same "normal" profit. While bureaucrats do not appreciate conflict among people, it adjusts the stock of a phenomenon to saturation more efficiently than government planning.

Section 13

13.1 The Law of Price Adjustment

u(s), μ and σ change over time for a variety of reasons, each change necessitating a recalculation of the saturation point. It is the business of entrepreneurs to anticipate these changes and to adjust stocks accordingly. While most shifts in a demand distribution are of only local concern, one is of particular interest to economics. If some of the people represented by the demand distribution for a phenomenon receive money from the government, how does the saturation point change? Whether these people receive a grant, a low interest loan, or are doing contract work for the government, they are more liquid than they want to be. Knowing the negative effect of a loose monetary policy on the value of money, they are not going to hoard it. Relative to money, the importance of phenomena has increased. How are prices and stocks affected and which adjusts more dramatically to the increase in μ?

It is an old adage that people get more out of something the more they put into it and, money being the measure of utility, one expects increases in the importance of a phenomenon relative to money to increase the phenomenon's price in proportion to the price that it has already attained. Mathematically, $p = p_0 e^{\mu}$, with p_0 the price at saturation with no importance relative to money and p the price such that $f(\mu,m) = S(\mu,m) - mc(\mu,m) = 0$. Notice that p is the particular price which satisfies the condition $f(\mu,m) = 0$ while m denotes an arbitrary price. Variables included in the functional notation are allowed to vary while others which appear in a function but are not listed in the parenthesis of the function are assumed to be constant. Here, we are discussing changes in both price and importance where before only price was allowed to vary.

Theorem 12: The price at saturation increases exponentially in response to an increase in the importance of a phenomenon relative to money; that is, $\frac{dp}{d\mu} = p$.

Proof: $f(\mu,m) = S(\mu,m) - mc_0(\mu,m) = 0$ implicitly defines a level set in the μ,m plane. Let that level set be parametized by $[\ \mu(t)\ \ m(t)\ ]$. By the chain rule, the derivative of $f(\mu,m) = 0$ is $\frac{\partial f}{\partial \mu}\frac{d\mu}{dt} + \frac{\partial f}{\partial m}\frac{dm}{dt} = 0$ or $\begin{bmatrix} \frac{\partial f}{\partial \mu} & \frac{\partial f}{\partial m} \end{bmatrix} \begin{bmatrix} \frac{d\mu}{dt} & \frac{dm}{dt} \end{bmatrix} = 0$. The latter vector is the derivative (tangent) of the parametized level set, so $\begin{bmatrix} \frac{\partial f}{\partial \mu} & \frac{\partial f}{\partial m} \end{bmatrix}$ is perpendicular to the level set which passes through any μ,m where it is evaluated. From the definition of saturation, this is downward (toward smaller m), so a 90° counter-clockwise rotation of $\begin{bmatrix} \frac{\partial f}{\partial \mu} & \frac{\partial f}{\partial m} \end{bmatrix}$ is tangent to the level set of all μ,m combinations with $f(\mu,m)$ constant. Dividing its vertical component by its horizontal component gives the desired rate of change in price. $\frac{\partial f}{\partial m}$ and $\frac{\partial f}{\partial \mu}$ are given in Theorems 8 and 9, respectively.

$$\frac{dm}{d\mu} = -\frac{\frac{\partial f}{\partial \mu}}{\frac{\partial f}{\partial m}} = \frac{m\frac{\partial f}{\partial m}}{\frac{\partial f}{\partial m}} = m$$

$$\text{with} \quad \frac{\partial f}{\partial m} = \sum_{r=0}^{\infty} c_0(x)\left(\frac{\ln(x)-\mu}{\sigma} - 1\right) \quad \text{and} \quad x = \frac{mu(0)}{u(r)}$$

This relation is true regarding the level set which passes through any point μ,m. Choosing only points along the level set $f(\mu,m) = 0$ (rather than another constant) yields $\frac{dp}{d\mu} = p$. ■

Notice that $f(m)$ in the above proof may be expressed as

$$f(m) = -\int_m^{\infty} f'(t)dt = \int_m^{\infty} \sum_{r=0}^{\infty} c_0(x)\left(1 - \frac{\ln(x)-\mu}{\sigma^2}\right)dt \qquad \text{with } x = \frac{tu(0)}{u(r)}$$

Incidentally, it does not matter that the rotation is counter-clockwise since a clockwise rotation also switches the components but negates $\frac{\partial f}{\partial \mu}$ instead of $\frac{\partial f}{\partial m}$. Because the sign comes out front after the division, it is immaterial which way $\left[\frac{\partial f}{\partial \mu}\ \frac{\partial f}{\partial m}\right]$ is rotated.

An alternative proof uses the chain rule to differentiate $f(\mu,g(\mu,m)) = 0$ with $p = g(\mu,m)$ to get

$$f_\mu(\mu,g(\mu,m)) + f_m(\mu,g(\mu,m))g_\mu(\mu,m) = 0$$

This equation is solved for $\frac{dp}{d\mu} = g_\mu(\mu,m)$. For $f:\Re^{1+1}\rightarrow\Re^1$, however, recourse to the tangent seems more intuitive than a purely algebraic proof and the former was chosen as the main proof. A more detailed discussion of implicit differentiation appears in Appendix C regarding another problem.

By Theorem 12, the price at saturation increases exponentially in response to an increase in the importance of a phenomenon relative to money. What about stock? Intuitively, one expects stock to remain constant since, effectively, all the government does by issuing money is to change the figures in which prices are quoted and that should not affect the stock of phenomena that people keep in existence. Most economists would agree that this is true in the long run but would argue that, because of the uneven diffusion of fresh issues of money, the stock of phenomena is temporarily affected. Money diffuses unevenly from a central bank and that is the principal motivation for issuing it (otherwise those close to a government would not profit from their connections), but I assert that this does not provide any incentive for the stock of phenomena to increase.

Even in the framework of mainstream microeconomics it has been difficult for economists to justify government issues of money, requiring first a "new" and then a "new-new" microeconomic theory, both of which were unsatisfactory. The motivation for economists to insist that the stock of phenomena is increased by a government issue of money is twofold. First and most obvious, they live on government grants and like to feel that this action has a redeeming social value, namely that it increases the stock of that which they spend it on. Second, they may be confused as to what wealth is. The wealth of a person (or community) is his (or their) stock of phenomena, not how much he (or they) may produce in a given time period. There are always fluctuations in production, especially during eventful times such as a fresh issue of money from the government, but these fluctuations are small and largely cancel each other out, being mainly caused by errors in forecasting. It will now be proven that government expenditures provide no incentive for the stock of phenomena (wealth) to increase. In the next chapter, it will be shown that depressions are caused by the wastage which accompanies all government activity.

Theorem 13: The stock at saturation remains constant in response to an increase in the importance of a phenomenon relative to money; that is, $\dfrac{dS_p}{d\mu} = 0$.

Here, the subscript on stock denotes that it is the stock associated with the saturation price, p.

Proof: We are interested in the change in stock along the level set implicitly defined in the μ,m plane by the relation $f(\mu,m) = 0$. As noted in the preceding proof, the tangent to this curve is $[\ 1 \quad m\ ]$. Normalizing this vector and taking the inner-product with the derivative of $S(\mu,m)$ gives the desired rate of change in stock. Since we are interested in proving that this change is always zero, it is sufficient to show that the numerator is always zero and we may omit normalizing the directional vector. The inner product of this with the derivative of stock, $\left[\; m\sum_{r=0}^{\infty} c_0(\mu,m) \quad -\sum_{r=0}^{\infty} c_0(\mu,m) \;\right]$, is zero. ■

Together, the two preceding theorems will be referred to as the Law of Price Adjustment. A graph of $p(\mu)$ and $S_p(\mu)$ is given in figure 3. Because Theorem 13 is a corollary of Theorem 12, the term "Law of Price Adjustment" is used to denote both theorems. From a practical point of view, however, the assertion that the stock of phenomena is unaffected by depreciating a currency is more important because, by definition (see section 8.3), economics is concerned with the wealth of a nation. Of course, the wealth of an individual can always be increased at the expense of other people by printing and spending money, but theoretical economics (hopefully) addresses more lofty aims.

It is important that the Law of Price Adjustment does not place any restrictions on marginal utility, on the importance of a phenomenon relative to money, or on the difficulty of substituting other phenomena. Within my economic theory, these three characteristics are all that distinguish phenomena from one another; that is, phenomena with the same $u(s)$, μ, and σ are isomorphic. Thus, it is impossible to argue that my theory is inapplicable in certain situations because it has been proven to apply to all possible situations; that is, it applies to phenomena at every point in $u(s),\mu,\sigma$ space. Since any mathematician will confirm the deduction of the Law of Price Adjustment from the three axioms (listed in section 11.3), for an economist to accept or reject the Law of Price Adjustment is equivalent to his acceptance or rejection of the three axioms, respectively. Attempts to divert the argument away from the acceptance or rejection of the theory's axioms should be discouraged.

The implications of the Law of Price Adjustment should be obvious to anyone who has studied mainstream economics; stickiness of prices is the cornerstone of Keynesian Economics. Even for those who do not follow the mathematics, common sense alone is sufficient to refute the Keynesian premise. Considering that a government can print money for itself within a day's notice, if the adjustment process could not be done in equal time, the whole system of indirect exchange would have collapsed long ago. Prices can be changed with a word, but the stock of phenomena can only be changed after considerable toil. It is obvious which is adjusted and which left constant.

Having arrived at a position so fundamentally opposed to mainstream economics, it is important to realize exactly where we parted company. The difference is that my theory is concerned with the price and stock of phenomena while mainstream economics is concerned with the price and supply of phenomena. I assert that the stock of phenomena is more important than the supply because all of the decisions made regarding a phenomenon are based on its stock (how much of it is in existence), and not on how much of it happened to be produced in some arbitrary time period. Phenomena are the same whether they are produced in one time period or another. Most people do not know and none care what the supply of phenomena is, they are concerned with the stock; this week's or month's supply is only a small part of the available stock. Even if a factory is temporarily closed for a week or a month, the price of its product is hardly affected because the total amount of phenomena in existence is hardly affected. Yet during that week or month the supply is zero. Mainstream economics, which relates price to supply, is unable to explain why the price does not increase dramatically as inspection of the supply and demand curves predicts that it should.

Parking on campus has a price, so mainstream economists must believe that there is a supply, that is, an influx, of parking spaces. Yet none are being produced. Clearly, it is the stock, the absolute quantity of them, that determines price. Supply never means anything in economics, though sometimes (for non-durable phenomena) it can pass for stock. There are three principle mistakes of mainstream economics, but addressing supply and demand instead of price and stock is the most egregious. The other two are assuming that all short-term credit instruments function as money (this will be discussed in Chapter Five) and believing that the average price level is a meaningful statistic and, hence, that prices are "sticky."

Stock is a stable parameter, as the stock of most phenomena is large relative to their supply. Supply is very volatile. For instance, it is zero every night when the factory shuts down and yet, contrary to the logical conclusion of mainstream economics, day and night prices are the same. Actually, few economists take the term "supply" so literally, but many have noticed that their concept of equilibrium is a precarious one. **Axiomatic Theory of Economics** does not have this problem, as it is concerned with price and stock rather than supply and demand.

$u(s)$, μ, and σ change over time for a variety of reasons and it is the business of entrepreneurs to anticipate these changes and to adjust stocks accordingly. That they are good at this is evidenced by the wealth of the nation. Actually, most would-be entrepreneurs are very bad at this, but they have only their own fortunes to lose and generally do not pull anyone else down with them. They are expendable. Capitalism does not provide an invisible hand to guide entrepreneurs, but it does provide an invisible foot to kick out the inept ones. (Here I am assuming a free banking system - low quality loans are the hallmark of a central bank.) It is in a command economy that the problems of time and ignorance become most important, for the ignorant often aspire to the purple and, once in power, they can pull the entire economy down. Command economies only work when there is a clear and present danger, like an advancing army that needs to be repelled. In peacetime, government intervention in the economy can only result in despair and ruin.

The extension of stock to a national economy is the wealth of the nation and the extension of supply to a national economy is Gross National Product. It is well known that mainstream economics prescribes war as a remedy for recession; that is, they assert that war increases Gross National Product. War does increase the Gross National Product for the simple reason that the income of military contractors, like everyone's income, contributes directly to it. But the increase in Gross National Product is accomplished only by liquidating the nation's wealth and I assert that it is a nation's wealth that is important, not its Gross National Product. War destroys wealth and from an economic point of view, that is all that needs to be known about it. From a non-economic point of view, some people find war exhilarating and some find it horrifying. It is not the purpose of this book to comment on that debate, but it is the purpose of this book to assert that war is bad for an economy.

13.2 Critique of the sticky-price paradigm

Why do mainstream economists think that prices are sticky if they are not? It is because economics has always been concerned only with the overall price level, that is, the average of all the prices in an economy. There are so many prices in an economy that the few markets patronized by people whose money holdings are inflated by the government cannot drastically affect the average. Eventually, new money diffuses throughout an economy and all prices rise slightly, but by that time nobody cares. People are concerned with relative price changes, not the average level of prices, most of which are for phenomena that they do not even buy. Individual dealers do not equate it with their specific costs; they know what their costs are by reading the figures on the checks they sign. The average price level in an economy is probably the most useless statistic economists could have invented.

The average level of prices is "sticky" because it takes time for money to diffuse through a community and if one is averaging all prices, it is some time before one notices a change in one's statistics. This average is also meaningless for the same reason. The effect of issuing money is to redistribute wealth to the people who receive the new money first and that is only possible because of the slow diffusion of money through an economy. If everybody were to add a price-change premium immediately upon hearing of a new issue of money, the average price level would not be sticky but neither would the people with the new money be able to buy as much. The assertion of mainstream economists that the average level of prices in an economy is a meaningful statistic has done more damage to their credibility than any other assertion they have made.

For reasons that will be explained in the next chapter, the existence of a central bank causes swings in the stock of money. Each change in the policy of a central bank produces distinct classes of winners and losers. Thus, at any given time, prices of different types of phenomena are moving in very definite but opposite directions. Concerning oneself with the average of such prices is ridiculous. Also, the same event can produce different results, depending on the attitude of the central bank and speculation about how its attitude will change in the future. For instance, when a central bank raises interest rates to protect its currency, it helps commercial banks at the expense of businesses, but only to a point, for, if businesses default on their loans, banks collapse too, as high interest rates are only profitable to banks if people are paying them. The panic of possible bank failures causes speculation that the central bank will dramatically increase the stock of money to bail out failing banks. In general, however, only large banks are completely insulated from risk. This creates a new set of winners and losers that depends, not only on their status as creditors or debtors, but also on their size. None of the prices involved in this process are sticky, but they are all moving in different directions and changing direction against the backdrop of a steady central bank policy. When central bank policies are themselves sporadic, as they inherently must be, efforts to average prices lose all meaning. As will be shown shortly, the average of prices for different phenomena is definitionally without meaning, but when a central bank has divided people into distinct classes of winners and losers and different prices are rapidly moving in different directions, their average is not even a useful approximation. Discussions of an average miss the whole point of monopolizing the issue of money, that is, division of society into winners and losers.

If two central banks are pursuing equally reckless policies, but the business cycles that they are creating are somewhat out of sync, one will fail before the other. Thus, even though their policies have been nearly identical, there is a short time while one community is in boom and the other is in bust. Businesses in the later community will have a fire sale as they liquidate everything they own in an attempt to save themselves. If the central bank in the former community considers the average price level to be a meaningful statistic, they will be delighted to see it go down because of their neighbor's fire sale. In a display of genuine ineptitude, they will redouble their credit expansion because they are encouraged that they can do so while the price level is down. It apparently does not seem important that it was a similar credit expansion that caused their neighbor to make the low-quality loans that are now in default. This is the type of egregious mistakes that are made by economists who attach meaning to average price level statistics.

Historically, most major crashes have been preceded by a stable or slightly falling average level of consumer prices. There are several reasons for this:

1) The crash would not qualify as "major" if the central bank was not encouraged by low consumer prices. Thus, for whatever reason that they are low, this is the principle reason that minor crashes become major ones.

2) Credit expansions affect the prices of corporate stock and capital more than of phenomena that is ready for consumption. The low interest rates encourage companies to embark on grandiose capital projects in spite of the relatively low prices that their final product is selling for.

3) When the illusions of the boom are revealed, the first companies to go bankrupt dump their inventories, temporarily depressing consumer prices. This often occurs quietly while the big companies (the ones close to the government) are still booming and so the average price level declines before the economy-wide crash.

Labor unions are known to write contracts in which their employer is given a free hand in setting the employment level as long as he pays according to the wage schedule. The only restriction on layoffs is that they respect seniority. If a fixed percentage of every laborer's wage goes to union dues, the union's wealth is proportional to aggregate utility. However, by the Law of Price Adjustment, increases in the importance of a phenomenon relative to money cause the price to increase exponentially while stock remains constant. Thus, if the contract is going to be binding for any length of time, unexpected fluctuations in demand are less likely to hurt aggregate utility if the stock is fixed rather than the wage schedule. In fact, the owner of a plant would not do badly if he always ran it at an efficient level and simply had sales whenever his warehouse became crowded and raised his price whenever he experienced difficulties meeting orders on time. This is somewhat the principle already applied by contractors who bid on each job they do and who attempt to do several jobs at once; there is an inherent variance in the price of custom jobs anyway. Unless the owner is very close to the government, he does not experience any large swings in demand which cause this pricing policy to appear sporadic, and if he is close enough, he can assure his position despite changes in their economic policy. A steady hand on the level of output prevents those only moderately close to the government from being induced into expanding their plant only to find that the next administration has other plans for them.

Another reason for fixing the employment level rather than the wage schedule is that laborers can take a cut in pay if they have to, but getting laid off so that their comrades can maintain a high standard of living causes resentment. Laid-off union members with fixed debts will abandon the union rather than wait for a recall. Unemployed people do not appreciate hearing about union workers who are making several times more than they are worth per hour. Actually, not very many people can do that consistently. The world is full of people who are quick to talk about the huge sums of money that they make per hour but, when pressed, admit that they only get that work once in a while. Most workingmen understand that unions are not helping them find work, though many of them never quite abandon the vague hope of getting one of those high hourly wages that they know exist. Also, some hear about "the interests of labor" and think "When I get work, it's labor, so that must be me" and then they blindly support the unions.

From an employer's point of view, a stable employment level is more efficient than having his plant run full blast one month and be idle the next. Also, many employers resent the seniority system; they would like to be able to fire anyone who angers them. With a contract fixing the employment level, they could, the only requirement being that they fill the position immediately from a list provided by the union rather than from the general public. Most union workers are hardworking but a few bad apples with seniority have given them all a reputation for laziness. The proposed system allows unions to clear their reputations by setting up guilds which train young workers and eject incompetent ones. Employers would not resent hiring from such lists because they have long known that it is easy to be deceived by pretenders when hiring from the general public. This would be an improvement over the current system of trade schools set up to fleece unemployed people who are induced by the availability of student loans into paying exorbitant fees for meaningless diplomas. Thus, there is no necessity for contracts fixing the wage schedule, which might cause some price stickiness if they were to become widespread. Historically, however, they have never been that prevalent.

Retailers are known to set their prices at a fixed markup over costs. This is a wise policy for a retailer because he deals in so many products that he cannot be familiar with the market for each one. Thus, he defers the pricing decision to the dealer, who should be familiar with the market since it is the only thing that he sells. This policy also insulates the retailer from accusations of price gouging. Since he is not really selling the product so much as the service of bringing it to consumers, he does not care if the dealer's pricing decision is wise or not. That is the dealer's problem; the retailer can always drop an unprofitable product.

Does this policy cause prices to be sticky? No, it defers the pricing decision to exactly the people who try to set it at saturation, and that is what the Law of Price Adjustment addresses. There may be local runs on some phenomena which do not affect a dealer's pricing decision because he is analyzing the world market. But it is too much to expect such local shocks to be fully accounted for at the shopkeeper's level. A retailer does his best in deciding which items to put on sale, but he may stock thousands of different products and cannot analyze the market for every one. Actually, the price indexes currently in use do not ask about sale prices, only "regular" prices, which is surprising, since those indexes are the only empirical evidence given for price stickiness. Most retailers advertise a new list of sale items every week (witness the junk mail in your mailbox every day) and, as is well known to consumers, spot checks on prices depend largely on which retailer one talks to. Furthermore, the great diversity of superficially-similar products of varying degrees of quality and convenience make such polls virtually meaningless at the macroeconomic level. Anyone who has undertaken a comparative poll of prices charged in his area for a phenomenon knows how little is learned until he visits the stores to verify the advertised characteristics of their merchandise. How ludicrous it is to average the results of such a poll, or worse, of one asking about many different kinds of phenomena, and then compare that average with the average of another poll taken years later!

Such an average is not just ludicrous but it is definitionally without meaning, for one need only ask in what units the result is expressed and one has found a contradiction. It is a common practice for economists to clothe their arguments in obscure statistical jargon, but even if the math truly exceeds the reader's comprehension, he can at least check the units. Tracing the units through an equation is not difficult and, if the reader finds that the result speaks of one thing but is measured in units from another scale, he need read no farther. Now, in what units does one express the average (summation divided by the number of observations) of the prices of several different kinds of phenomena? One can only sum up numbers denominated in the same units, that is, in money per quantity of a phenomenon. Even in the individual categories, the phenomena are not similar enough to have a uniform price and, if the categories are narrowed any farther, that is just the raw data, which was known already. Furthermore, even in a relatively homogeneous category, for what size of unit should one inquire about the price? The larger the unit the more effect it has on the overall average and also the more likely that its price is subject to quantity discounts. Thus, besides the complete lack of a scientific basis for weighting the importance of different prices on what is loosely referred to as the "price level", mainstream economists are doing the math wrong. Relative price changes are ratios and the representative of a set of ratios is their geometric mean, not their arithmetic mean. Economists should find the ratio of each phenomenon's current price with its price in a base year, multiply all of these, and then take the n'th root, with n the number of distinct phenomena. This avoids the internal inconsistency of the current method but it still does not address the weighting problem. Hence, the economy-wide price level is a meaningless statistic whose units make no sense. The units problem can be corrected, but the meaninglessness cannot.

One may still use non-parametric techniques but, by definition, that means one cannot attach a magnitude to the change in the price level but only a direction, which is not very useful. Parametric techniques (such as taking an average) are only applicable to data from the same metric space; otherwise one can only attach a sign to the data as having increased or decreased but cannot compare the magnitude of their changes. But even then, technology changes so much from one survey to the next that the result is virtually meaningless. If economists cannot even state with confidence whether the price level has gone up or down, they should consider admitting that theirs is not an empirical science but an abstract one.

13.3 The definition of the term "rent"

First, let us clarify what is meant by the stock and price of phenomena, labor in particular. The stock of laborers with a particular skill, like the stock of any phenomenon, is a static concept. It means, at a specific time, how many are working as opposed to being idle or working in another field. The interval between measurements should be made as narrow as the employers' measurement and then averaged over the pay period. If the wage is specified by the hour and workers receive their paychecks every week, economists should poll employers every hour for a week and then average the results. This difficulty in measurement is associated with any rental price. The stock associated with the sale price of a commodity whose ownership changes with each transaction is easier to measure; it only requires one poll. While economists have built complicated theories about rent as opposed to sales, nothing distinguishes rent from sales except difficulties in measuring stock. It is important to realize that the stock associated with a rental or a sale price is always an absolute quantity, not a rate. This absolute quantity applies to a specific temporal interval in history. For labor, the interval is one week long, while for phenomena whose ownership changes with each transaction, the interval is instantaneous.

Rent is the price charged for using a phenomenon for a certain time period and then returning it to its owner. Many laborers receive a monthly salary, especially the more skilled ones, and because even the least skilled require training and it is the custom not to fire people without notice, it can be assumed that all employment decisions are made on a month-to-month basis. Rent for land, buildings, and machinery are also made on a monthly basis. Open accounts with suppliers or with credit card companies are not charged interest if paid within the month, the price being raised to include a month's interest. Therefore, from here on, all prices will be normalized to their monthly rent. Interest and price compensate for each other. A dealer may advertise a zero interest rate and then boost the price or he may advertise a low price and make up for it with a high interest rate. In fact, the decision is usually left to the buyer as to whether he wants a rebate or low interest financing; prices can hardly be more flexible than that. There is a lot of variance in prices and finance terms, especially when they can be adjusted at the discretion of individual buyers, as they can when a dealer offers his own financing. This is the main reason for normalizing all prices to their monthly rent; it is more representative of the value of phenomena relative to other phenomena. In the next chapter, a less local cause of instability in interest rates will be discussed which will make the concept of rent particularly useful.

Putting all purchases in the same time frame also helps people work out a budget, especially laborers who live from paycheck to paycheck. The wealth of someone living like this is equivalent to his income, but all of his decisions to consume or save are still made out of his wealth and it is wealth that measures how well he is doing. This is obvious for people with assets; they have to make decisions about everything they own, not just what they happened to earn during the current month. It should be equally obvious to everybody, and it was a grievous mistake of mainstream economists to make income central to their theory. Economics is defined as being concerned with the creation of wealth; the actual creation of wealth can only be damaged when theoretical economists neglect this point. That wealth is indeed neglected by mainstream economics is evidenced by the fact that advanced courses in macroeconomics are called "income theory" and it is income, not wealth, on the horizontal axis of AS-AD and IS-LM graphs. The price level, which I have already asserted is meaningless, is on the vertical axis of AS-AD graphs. The interest rate is on the vertical axis of IS-LM graphs. See Appendix C. Because price adjustments compensate for changes in the interest rate, often at the discretion of each individual buyer, it should already be clear that I do not find interest rates very meaningful either, at least not in the context of macroeconomics. Of course, many people have found mainstream economics to be meaningless, but they have nothing better to offer. Hopefully, the theory presented in the next chapter will be more meaningful to them. Engineering students may even enjoy filling their school's soft science requirement.

CHAPTER V

MONEY AND CREDIT

The curious view that "probability runs in streaks"... brings to mind how so many, even well regarded, pundits of subjects such as gambling, sports, economics, etc. confuse their own verbal reaction to a past event with an actual explanation of it and an augur for the future. Thus we have the... stock market analyst who cannot report a fall in price without conjuring up "selling pressure."

Peter Griffin
The Theory of Blackjack[13]

Section 14

14.1 Introductory remarks on the spectrum of liquidity

The more liquid members of a complementary set do not always seem to get their full share of the value of the set and, as production is often controlled by the owners of the less liquid members, it is these owners who are blamed. Whether or not labor or other liquid phenomena have saturated their market and should consider excluding some of the stock is a subject that has already been taken up. Concerning exploitation, however, it must be remembered that putting the several members of a set together takes time and the resulting set is itself a subset of larger sets. Thus, from the employment of the liquid members in question to the consumption of the maximal set, a good deal of time passes. Liquid members who are unable to wait for the realization of their contribution pay for this weakness by accepting less than it is worth upon completion.

As there are few people with the wealth to advance the capital needed for long, productive processes and those with a desire to invest their wealth rarely have the time or the inclination to familiarize themselves with various productive processes or the ability to forcefully pursue bad debts, banks have been organized to intermediate. For taking on risks and analyzing various investment opportunities, they are able to loan capital at a higher rate of interest than they pay to borrow it. Banks and the phenomena of interest rates which accompany them are a convenience to savers and investors but they do not preclude directly investing one's labor in the production of capital. Wealth has to be created and, while financial institutions may facilitate that process, they cannot alter its physical parameters.

Bank loans are always in the form of money which an investor uses to buy the particular capital that he needs. This has caused mainstream economics to equate capital with money and to assert that increasing the stock of money also increases the stock of capital and thus of wealth. Money and capital are at opposite ends of the spectrum of liquidity; money is the most liquid of all phenomena and capital is, by definition, non-liquid phenomena because it represents a commitment to future production. To increase the proportion of liquid phenomena in one's wealth implies decreasing the proportion of capital in that wealth. Increases in the stock of money rearrange the distribution of capital toward the people who control the issue of money. However, because their use of this capital is easier to see than uses that only might have occurred in different circumstances, mainstream economists think that the stock of capital has increased.

Actually, capital is only accumulated by the ability of people to wait for a return on their wealth rather than consuming it at once, that is, their ability to save. Saving and investment are equivalent except for the savings that a government borrows or confiscates, which they consume. The identity of saving and investment is tautological. It is independent of wealth or income. Investment is determined by the incentives for the stock of capital, like all phenomena, to move toward saturation. Because investments in capital often require bank loans, however, there is a weak link in the process where a government may borrow or confiscate people's savings. Capital has not always been so vulnerable. When agriculture and cottage industries were predominant, people saved simply by investing their labor in land and capital improvements, which never passed though an intermediary as still-liquid phenomena which could be used for government consumption. Also, when people employed the various commodity moneys convenient to farmers and tradesmen, they were largely immune to an unexpected depreciation of their monetary unit.

14.2 The relation between rent and interest rates

Rent is what an investor foregoes to acquire capital. In primitive economies, this means foregoing the leisure one normally enjoys after the day's needs are satisfied. It does not take long to satisfy immediate needs but, primitive or advanced, there is no end to the useful things that can be done with capital, as one never exhausts one's dreams. With the introduction of credit markets, employees can still work on their own projects, but they can also buy corporate bonds or loan some of their money to banks to be passed on at interest to investors for the purchase of capital and the hiring of people to produce capital. If investment and the invention of new methods exceed the decay and obsolescence of capital, it is accumulating. This is a slow process and it is difficult to measure exactly. In the short run, the stock of capital is constant and every unit of it has a constant rent. If one's savings are confiscated or eroded by depreciation of the monetary unit in which it is denominated, the stock of capital decreases and rent increases but, by definition, that is the long run. Investment fluctuates in short periods of time, but these fluctuations are small compared to the stock of capital and it is the stock, not its rate of change, that determines employment and output. The stock moves slowly toward saturation, usually upward, except when it is being wasted by the fiscal or monetary policy of a government.

With every new batch of money, interest rates fall and with every threat of a bank run, credit is tightened and interest rates rise. The stock of capital remains constant throughout, however, and likewise its rent. The only effect is that prices are high and interest rates low during credit expansions and, conversely, prices are low and interest rates high during credit contractions. Interest is only one component of rent. The other component, price, automatically adjusts to the interest rate so that rent remains constant in the short run. The stock of capital (or the rent; they are monotonically related) is important. It uniquely determines output. The interest rate, which is only one component of rent, is unimportant. Identifying rent with the rate of interest has caused a great deal of confusion during historical periods when a central bank has existed which could arbitrarily issue money.

The concept of a "natural" rate of interest, which the actual rate of interest (charged by banks) tends toward, is a trap that virtually every established school of economics has fallen into. The "natural" rate of interest is variously ascribed to be the ratio of prices for consumption and production phenomena, the rate determined when actual capital is lent without the intermediation of money, the rate that maintains the *status quo* (as measured by employment statistics), or just what people think is fair. The first assertion is meaningless in the aggregate (see section 13.2) and, for specific phenomena, it is what is usually defined to be the marginal efficiency of capital (profit), not the interest rate. This confusion has led to conclusions that are exactly contrary to the well-known fact that people borrow when their expected profits exceed a bank's interest rate (see Appendix C). The second assertion is meaningless because, as discussed in section 11.3, interest is a strictly monetary phenomenon and does not exist in a barter economy. The third assertion confuses the stock and the supply of capital. It is the stock of capital that uniquely determines (useful) employment because, both quantitatively and qualitatively, the amount of available work is determined by the amount of capital that has not been wasted and, at every instant of time, people are fully employed at the available work. Issuing money to loan at a low rate of interest is a negative influence on employment because it wastes capital and is only a positive influence when the money is given as a charity to unemployed laborers in "make-work" programs, but that is not useful employment. The fourth assertion, which is at the heart of most discussions of a "natural" rate of interest, is not even in the realm of theoretical economics.

However the "natural" rate of interest is defined, if it is defined at all, the concept has led to some of the most serious mistakes in economics. For instance, it has sometimes been asserted that, in the event of an unusually low interest rate, people hoard money in anticipation of loaning it after interest rates have returned to the "natural" rate. If they do, they will be disappointed, because a central bank can hold interest rates down indefinitely by issuing money. There are always borrowers because people want to borrow money to purchase capital as a hedge against depreciation of their monetary unit. Rent is rising and they want to get whatever capital they can with borrowed money, often capital they cannot use. They know that by the time their loan matures, the money they pay back to the bank will be worth less. The only limit to the issue of money is political pressure to stop depreciating the currency. By the time the issue of money gets that severe, speculative hoards will be next to worthless. There are no speculative hoards of money; in fact, it is the rush to buy capital which is responsible for its wastage during the boom. It falls into the hands of people chosen only because they were the first to get newly-issued money. Thinking that they can always get more, they commit it to grandiose schemes. The stock of capital is constant in the short run and it is not increasing in the long run because of wastage. Capital is also being wasted because people do not add enough of a price-change premium to their prices. What seems like unusually brisk business is actually the result of unwittingly making a gift of part of one's capital to consumers.

A central bank causes capital to be wasted because it makes it appear that there is more capital than there actually is. Bankers and entrepreneurs convince each other of the soundness of outlandish schemes and futuristic projects that would only be justified in a much wealthier society. Loans are made on agricultural land that can only be repaid if the price of food increases dramatically. Industries are built in less-developed countries that, without further loans, cannot even maintain the machines when they break down. In general, loan quality deteriorates dramatically in the presence of a central bank. The problem is not overinvestment. Overinvestment is impossible because the stock of capital should always be (and is) fully employed. The problem is that it is employed badly. When ill-conceived projects are abandoned, the capital in them may appear idle, but it is not. It is lost.

Eventually, the stock of capital falls so far below saturation that people invest their labor directly in capital production. This is inconvenient and hampers saving because of its inconvenience. Nevertheless, the rate of saving is determined by the stock of capital's relation to saturation. The rate of interest, which is set by the central bank, does not determine the rate of saving, only whether saving is done conveniently through the intermediary of banks or inconveniently through direct investment of labor in capital. When banks fail, it appears that investment has dropped as people turn to the direct investment of their labor in capital production. For this reason, mainstream economists believe that investment is procyclic. It is actually anticyclic. What appears to be investment during a boom is mostly wasted and what appears to be unemployment during a depression is really capital accumulation. It is not a very exotic capital accumulation, though, as it is primarily directed toward the production of food and shelter and only later to the luxuries that seemed so important during the boom.

14.3 The effect of government spending on national wealth

The accumulation of capital is a long-running process and hence, in the short run, the stock of capital is constant and government expenditures are taken directly out of private investment. The easiest way to answer economic questions is to look through all of the graphs and equations and see a fixed quantity of capital and determine how it is to be divided. Then, despite concepts like "an elastic money market", the answer is the same: Good for the people with the capital, bad for the people without. There is no idle capital waiting to replace what is consumed by a government, whether it is through fiscal or monetary means that they acquire capital.

A government has two methods to pay for its consumption: It can tax people or it can sell bonds. Taxes are just confiscation and are obviously taken out of private investment. Bonds that mature in the hands of private parties (not the central bank) compete with corporate bonds. Because the buyers hold them to maturity, they are interested only in receiving interest and would buy corporate bonds if the treasury did not offer them a higher interest rate. Thus, the two forms of fiscal policy, taxing and borrowing from the public, either confiscate or crowd out an equal amount of private investment. Bonds sold to the central bank (either directly or to private dealers who pass them on to the central bank before maturity) are paid for with a check that the central bank writes against itself. The treasury then deposits this check in a commercial bank, allowing commercial banks to buy government bonds amounting to several times the amount that the central bank bought. (The exact multiple is determined by the reserve requirement - see section 16.3.) Thus, the two forms of monetary policy, selling bonds to the central bank and to commercial banks, are effectively just printing money and using it to buy capital away from private investors. Capital consumed by a government through fiscal and monetary policies was never idle but would otherwise have gone to the creation of wealth. Common sense alone is sufficient to verify such an obvious assertion.

People are never so wealthy that they can allow capital to lie idle and hence the economy is always at full output; that is, people are producing all that they can. Full output is not something that an economy tends toward in the long run, it is a reality at every instant of time. If there are involuntary surpluses of labor, it is because there is not enough capital to complement the available stock of labor, but the economy is still producing everything it can with the resources that it has. It is capital that determines output; laborers just run the machines. It may be that the illusions of a depreciating currency caused people to waste their capital in projects from which it cannot be recovered, but that capital is not idle, it is lost. In the aftermath of inflation-induced waste, output is less than its previous level, but there is still no idle capital. If anything, poverty drives people to inelegant methods and they use capital they once distained as being too coarse. The economy being at full output, further issues of money raise prices but do not affect unemployment, that being relieved only by the reaccumulation of capital.

During a boom, people think that there is more capital than there actually is and this causes them to waste what they have on projects which would only be practical in a much more advanced society. Also, if one is unfamiliar with the effects of a loose monetary policy on the value of money, one may think that one is selling one's phenomena for a reasonable price when, in fact, one is liquidating part of one's capital to make sales. Depression-era landscapes have always been littered with the wrecks of futuristic projects that seemed reasonable during the boom but are more than a little absurd to the wandering unemployed during the depression. These projects may someday be redeemed after the country recovers from the depression, but that does not mean that capital is idle during depressions. Such projects are lost on the people during depressions because they are interested only in food and shelter. The problem with easy credit programs is that they prevent a country from addressing its highest priorities first.

I assert that, during boom times, capital is being wasted by every company, from the largest multinational down to the smallest mom-and-pop outfit. This is a harsh assessment of their business practices and the directors of small businesses might complain that I am being unfair to them. Small businesses cannot get credit at any interest rate, even during boom times, and are rarely seen building the grandiose projects that large businesses embark on during a boom. If they are large enough to be publicly traded, however, then some of their stock is held by people or organizations that can borrow from banks. Also, if the rest of the stock market is overvalued, companies will be pressured into reporting unrealistic profits to keep up. Slow, steady growth is unacceptable. Those who do not report extraordinary profits fail. Thus, during boom times, every quarterly report is treated as an emergency and companies consume their capital to meet that emergency. Because of this, the first sign of a bust is when small companies that had previously reported extraordinary profits start to fail and are suddenly discovered to be operating on a shoestring with worn-out and obsolete capital. Unfortunately, this usually goes unnoticed in the glow of the still rising stock indexes, which measure the value of large companies. Companies that are too small to be publicly traded also waste capital. Their owners do not reinvest their profits but let their businesses deteriorate while they buy stocks of larger, publicly traded companies. Thus, boom times are characterized by a transfer of capital from smaller companies to larger ones, and the big ones waste it. Also, while small businesses cannot get credit during a boom, home loans and car loans are easy to obtain, so a lot of capital is converted directly into consumer goods.

Recent history does not report swings in national income as severe as those in the more distant past, which has led many economists to claim that their policies have controlled the business cycle. This is difficult to measure, however, because people get used to stagnation and, as they forget their expectations, it is easy to redefine "potential" downwards so the statistics look better. In the past, most bank loans were to private parties, so booms and busts were characterized by fluctuations in private enterprises. People are quick to think that easy credit indicates a level of national wealth that justifies expansion of their business and they are just as quick, when credit tightens, to realize that it does not. Thus, when a government pursues an easy credit policy, national income increases dramatically and, when they tighten credit to bolster their depreciating currency, national income decreases as people liquidate their overoptimistic investments. Gradually, however, credit transactions in most countries narrow to the central bank loaning money to the treasury, so small businessmen hardly know what the interest rate is as they cannot get credit in good times or bad. Government waste is so institutionalized, however, that it does not fluctuate with interest rates, which is why economists now speak of stagflations and recessions rather than booms and busts, respectively. An extreme example of this is communist countries, which do not loan any money at all to private parties, and have replaced their business cycle with a permanent depression.

Sometimes a central bank loans money to the treasury of foreign governments, which raises a number of contentious but mostly irrelevant issues about foreign aid. Suffice it to say that such governments would not have credit if they were not acting as puppets of the creditor nation. People sometimes express wonder at banks that continually roll over loans to foreign governments without, apparently, expecting to ever get the principle back. They neglect to notice that the banks do the same thing with their own government. I assert that government consumption is wasted both at home and abroad, however, so the term "government" should include puppet governments as well.

While government consumption may serve important purposes such as defense, it does not directly produce wealth. Government waste is especially acute where the president does not have a line-item veto of legislation, though in any moderately-sized country, the bureaucracy is largely autonomous of the president regardless of his veto power. In most cases, he can only control the amount of government expenditures but not their composition. In general, a president can be confident that most of the spending decisions that he delegates are skimmed or go to "administration" without ever reaching the people that they are intended for. One can assume that by the amount which government expenditures exceed taxes, that much of national income is financed by liquidating the nation's wealth. Certainly, the people of a belligerent nation or of a welfare society would more quickly tire of their war or of their charities were its true cost laid out before them in the form of a weapons tax or a welfare tax. Also, taxes have the virtue of allowing discrimination in their application, say, away from the poor. The only virtue of financing expenditures with inflation (the central bank buying government bonds) is that it deceives people who are not familiar with economic theory. That there are many such people is easily verified by polling one's associates regarding who the national debt is owed too. The fact that none of them know demonstrates that the government's deception is working. All "great societies" and "new world orders" are built on this deception, for people would rebel if asked to pay taxes for such boondoggles.

Anyone who has dealt with a government bureaucracy can attest to the wastage involved, though this is an assertion that cannot be proven to one who does not want to believe it. In fact, the only bureaucracies that most economists deal with are those which distribute research grants, and the money is usually forthcoming for appropriate research. The reader must find examples of government waste himself;

it is unbecoming of an abstract treatise to dwell on examples of government waste because any event is dated and can always be rectified by legislation designed to remove it as an example. A theory that does not assume that most government expenditures are wasted, however, is not complete, for that situation is not precluded by the definition of government confiscation. It is my opinion that this situation is closer to the truth than the assumption of benevolence and foresight on the part of bureaucrats.

Paul Krugman writes[14] "Aside from defense and interest payments, the U.S. government is now mainly - yes, mainly - in the business of taxing the young and giving money to the old." By definition, retirees make no contribution to the economy and so, using Dr. Krugman's statistics, at least 43.9% of government expenditures are pure waste. Actually it is more, because most of the 13.7% spent on interest payments is for money borrowed and then given to old people. Thus, even without invoking the usual parade of welfare queens, architectural monstrosities, and hardware store items sold to the Pentagon at a several thousand percent markup, it is clear that at least half of the current U.S. government's budget is wasted.

Also, after a central bank tightens credit to save its currency, a large part of government expenditures go, not to the government's own boondoggles, but to failed businesses and their bankers. This is done to prevent the crisis of a bank run. However, such expenditures are, by definition, wasted, for sound businesses would not have defaulted on their loans. Thus, even the most liberal proponent of government benevolence and foresight must agree that this part of government expenditures is a waste. It is not a discretionary part of a government's budget, however, as a general run on banks is sufficient to topple almost any regime. The principle characteristic of a central bank is low-quality loans and, once a government has monopolized the issue of money, it cannot forego the expense of bailing out banks merely because it is wasteful. The government must either eliminate the monopoly on the issue of money or be content to look for waste in its own programs and grant the central bank as much as it needs to cover past defaults. Low-quality loans are inherent in the process of monopolizing the issue of money and can only be prevented with wholesale restrictions on loaning money to private businesses, which is referred to as a credit crunch. Since the central bank was created to finance the government, the tendency is to gradually eliminate all loans to private businesses, leaving only the government itself and a few semi-nationalized corporations, particularly defense contractors. Thus, by monopolizing the issue of money, a government can eventually achieve what could have been more easily accomplished, though less easily maintained, with the military, that is, communism.

Government waste is further compounded by overbudgeting in the world of high finance created by a central bank. Monetary matters conducted at the national level involve figures that are too large to be comprehended. The closest that most people come to high finance is buying or selling a house and, despite the volume of similar transactions conducted every day, the last three or four digits of the price are rounded off because it is impossible to estimate prices any more accurately. Central bank transactions are a hundred thousand times larger, so large that the value of money itself is affected by the deals. Furthermore, there is no precedent for such transactions because, by law, nobody else can issue money. The home buyer can use his salary and expected lifetime for estimating prices, but a central bank cannot do that since they do not have to earn the money, they just print it.

Thus, considering the immense influence that the government has on an economy, it should not seem anti-intuitive that changes in its policy can effect cycles of booms and busts. The roaring 20's are as well known for extravagance as the 1930's are known for hardship. It does not seem unreasonable that the capital wasted during and after World War I had something to do with the poverty of the Great Depression. This is an observation missed by mainstream economists who characterize booms as "recoveries" and busts as "exogenous shocks" and then study them a half cycle out of phase. People starve for lack of food and they are homeless for lack of houses, but the food and the houses did not just disappear, they were wasted. The problem that needed to be addressed during the Great Depression was reaccumulating capital. Most people would agree that building national parks was not as important as growing food or building houses. People may feel that theoretical economics is too academic, but when theoreticians forget what economics is about (the creation of wealth), there is no end to the misfortune and ruin that may befall a nation.

Figure 12

Capital Stock in the U.S. from 1930 to 1990

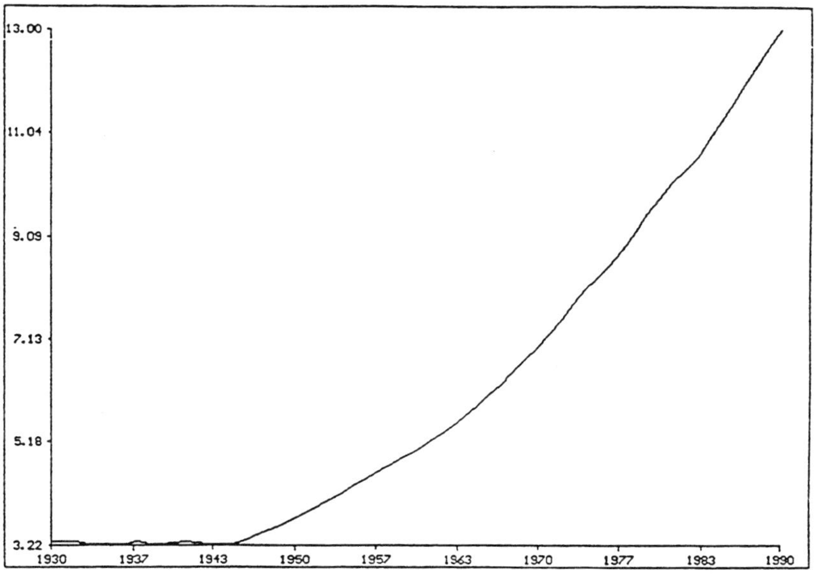

By the federal government's own estimate of the capital stock[15] (figure 12), it was almost constant from 1930 until 1947 and then increased steadily through 1990. The flat part is the depression and it extends through World War II, during which time civilians had fewer goods available than in the 1930's. This belies the assertion of mainstream economics that World War II rescued the United States from the Great Depression. By the definition of depression given in this book (a lack of wealth), the United States did not recover from the Great Depression until after the war. In fact, anyone who lived through the war can recall rationing and hardship, so it should seem strange that mainstream economists assert that war is good for an economy. War is always a great waste, not only of capital but of human lives, and it can only be justified if it prevents the greater loses that a foreign army would inflict were they to occupy the country. In 1936, John Maynard Keynes wrote in his **General Theory**[16]:

> *Pyramid building, earthquakes, even wars may serve to increase wealth.... If the treasury were to fill old bottles with banknotes, bury them at suitable depths in disused coal mines which are then filled up to the surface with town rubbish, and leave it to private enterprize... to dig the notes up again..., there need be no more unemployment and, with the help of the repercussions, the real income of the community, and its capital wealth also, would probably become a good deal greater than it actually is.*

Keynes was actually a hack, enlisted to justify the five year old and already entrenched WPA, which had people building national parks, not fighting wars. Since then, however, his theory has been used mostly to justify war. Of course, it would be simpleminded to assert that the United States only goes to war at the command of its economists; they do not have that much influence. However, it is not unreasonable to assert that the principle motivation of mainstream economics is to justify war. Since this is a book about economics, it is the economic motivations that concern us here. A complete study of the causes of war would have to address the interests of a number of people, not all of them on the side that initiates military action. It is unlikely that either the Japanese or the Germans were influenced by Keynes and it would be unfair to credit him with World War II, but neither is it entirely surprising that the United States went to war only five years after Keynes' call for more government spending. The Federal Reserve was created in 1913 to finance World War I, which went hot the following year, but until 1936 they lacked a theory to explain how war was good for an economy. World War I was, in fact, the first modern war. Here, the distinction between modern and ancient warfare is one of logistics, not weaponry. The only new weapon system introduced in World War I was the tank and that was a response to, not a cause of, trench warfare. The weapons that dominated the battlefield, heavy machine guns and artillery, were of better quality but conceptually no different from the gatling guns and cannons available during the American Civil War a half a century before. What had changed was that now they could fire millions of dollars worth of ammunition a day, month after month, for years at a time. (Also, they had trucks to transport munitions to the front, which would have been difficult with only horse-drawn wagons.) Historian Archer Jones observes that[17] "Only in comparatively modern times have states had available income taxes, fractional reserve banking, and debt monetization to divert a large portion of national output to support war."

National income (Gross National Product) is how many times money changes hands in a year, and that is indeed a positive function of government spending on weapons. The number of times money changes hands is increased by putting some in the hands of military contractors. However, I assert that the measure of an economy is wealth, not income, and that is only destroyed by a war, never created. Because of economists' concern for income instead of wealth, millions of people have died and untold quantities of money have been given to military contractors, much of it for nuclear weapons. Considering the danger of nuclear holocaust, it would have been safer to have just buried the money in coal mines as Keynes originally suggested. How anyone but a military contractor could think that fighting and killing is good for an economy is incomprehensible, but that is exactly the position of mainstream economics. Every one of their textbooks refers to the Keynesian Revolution as the beginning of modern economic thought, and anyone who has heard of the multiplier (it is introduced in first semester macroeconomics) knows how mainstream economists think government spending will affect national income. Appendix C discusses the multiplier in detail, but the essence of it can be learned in any macroeconomics course.

Statistics on employment do not distinguish between useful work which is actually creating wealth and busy work which is only slightly productive or is designed only to keep people off the street. During the Great Depression people in public works programs were, quite accurately, counted as unemployed. Today, government workers devote a great deal more effort toward appearing busy and many make-work programs are nominally in the private sector. If these people were correctly counted as unemployed and potentially skilled laborers in menial jobs were correctly counted as underemployed, the Great Depression would not seem so different. If our lifestyle has improved, it is not because we have reduced unemployment, but because of the amazing engineering feats that have taken place, especially in electronics. Economists are too quick to take credit for the wealth we have now compared to fifty years ago. We actually never recovered from the Great Depression, engineers just rescued us with technology. But people are still starving while the quantity of wealth wasted by the government increases every year. Estimates of wealth, however, are very vague because the value of capital does not lend itself to measurement. In the 1930's, a clever workman could reproduce almost everything he saw around him. Today, no homemade item can come close to what is produced commercially. In no previous time has science advanced to the point that the common man does not know how most of the technology around him works. It is ludicrous to compare events that are separated by half a century, particularly this half-century, simply by counting the number of laborers who answer "yes" to an employment survey. Who can say how things might have been if more attention had been given to the wastage of capital?

The greatest deception that mainstream economists have perpetrated on the general public is that employment statistics are a measure of how well they are doing. Many young people have never had a real job, have despaired of finding one, and have drifted into low-level crime. They view themselves as unemployed but are not counted as such because, without filling out employment applications, they never appear on surveys of job-seekers. Social workers will confirm that most of the people who seek their help are of the working poor, that is, people whose jobs pay only a fraction of their living expenses. Most of these people make ends meet with forays into crime, usually retail narcotics or prostitution. When caught, the resulting criminal record effectively prevents them from getting a real job for five to seven years or, if their career was in a sensitive area like health care, never. Whole industries have been organized around pressing people with criminal records into laboring in factories for half wages. Their lives are basically ruined, yet at no time did mainstream economics count them as anything but fully employed. The output of a laborer using modern equipment to do his job is many times greater than of one using primitive tools and, as it is his output that he is paid for, the quality of his employment depends on how well capitalized his industry is. That mainstream economists insist on measuring employment quantitatively but not qualitatively is a deliberate deception to conceal how much capital the government is wasting.

Technology is the root of all wealth; laborers just run the machines. The assumption that the labor force is the deciding factor in determining the wealth of a nation has long been a stumbling block for economists. No one can deny the anthropologist's observation that industrial nations tend to oppress people in undeveloped countries, but to assume that they derive their wealth from such exploitation is a mistake. The governments of industrial nations oppress undeveloped countries because they can, but the profit from it benefits only the government and its close associates. Many people who justifiably oppose imperialism have become communists because the social sciences insist on identifying imperialism with capitalism and anti-imperialism with communism. In fact, only the government and a small group of politically influential companies profit from the exploitation of people in a foreign country, or people of some race within their own country. Governments of capitalist and communist countries both employ secret police agencies which both use the same tactics (assassination, false arrest, forgery, anonymous allegations and innuendo) that offend anti-imperialists, but this is not the source of the country's wealth. The reason that only wealthy countries are associated with such actions is because only they have the tax base to support an intelligence community that can reach beyond national borders. Historically, no country, rich or poor, capitalist or communist, has succeeded in restraining its secret police within the rule of law.

The wealth of industrial countries is derived from technology. Technology has advanced more in the last half century than it has in all of the preceding history of mankind, though most of it has been wasted through misguided economic policies. Economists may decry growth as a long-term, background phenomenon beneath their consideration, but when one can see technological advances of historical proportions occur within one's own lifetime, that is not background information. Given the government's systematic wastage of capital, it is ridiculous to count the number of people with a job, however degrading it may be, and rejoice when the number climbs a fraction of a percent. The wealth of a nation is a much deeper concept than its quarterly employment figures. But, while it is easy to see the difference between creating wealth and wasting wealth, the absolute quantity of wealth is not amenable to measurement.

Thus, any quantitative statement about the wealth of a nation should not be taken very seriously. Quantitative statements are empirical and should not enter into the discussion of an abstract theory such as the one presented in this book. If economics is ever going to gain any respect, economists must make assertions based only on logic, not on statistics. Difficulties in defining and measuring concepts like GNP, the unemployment level, and the money supply are discussed in the beginning of every macroeconomics textbook but are then forgotten. The advanced student who rereads the introductory chapter of his income theory text where methods of measurement are explained will be amazed at the precise calculations that he has grown accustomed to making with those statistics. The only explanation for the use of such statistics is that, without them, economists would have nothing to write about. Economists consider themselves scientists, so they need something to be scientific about. The theory presented in this book is abstract and does not rely on the accuracy of empirical measurements. However, for economists who prefer to stay in the mainstream, I have developed a method for analyzing their empirical errors. It is discussed in the fourth part of Appendix C.

Section 15

15.1 The definition of the term "money certificate"

No commodity becomes a medium of exchange without being physically limited in quantity for, if it were not, people would simply produce more of it as the demand for it increases. The physical nature of commodity money, however, can cause technical difficulties in its use as a medium of exchange, primarily that it is hard to carry and cannot be traced if stolen. This prompts people to demand certificates for commodity money which are easier to use, such as banknotes or checking accounts. These will be referred to as money certificates or sometimes as demand deposits. Money certificates are receipts for commodity money that is deposited at a bank and which can be redeemed at any time. Banks take up this service in addition to their savings and loan business because people already trust them with their savings. Demand deposits are superficially similar to savings deposits because they both require trust. The difference is that savings depositors exchange present phenomena for future phenomena while the holders of money certificates only exchange one kind of present phenomenon for another. Demand deposits are a bailment; a money certificate is a receipt for phenomena held for safekeeping. As long as a bank of issue redeems its money certificates for commodity money whenever they are asked to do so, its money certificates are exactly equal in value to the corresponding commodity money.

For the purpose of economic theory, banknotes and checking accounts are equivalent. The proportion between banknotes and checking accounts fluctuates with the enforcement of laws that only affect one or the other. Some anti-inflation laws have been specifically directed against banknotes, thus promoting the use of checking accounts. Other laws that sanction the confiscation of people's checking accounts (criminals, tax evaders, debtors, etc.) may not have been meant to affect monetary policy but have tended to make banknotes more popular than checking accounts. Neither kind of money certificate is physically limited in quantity the way commodity money is, so it is up to depositors to pressure banks to redeem their money certificates on demand. The amount of pressure necessary for this, however, is well within the ability of the customers of a bank if they all cooperate with one another and are supported by their government. Historically, people have not always had this support from their government, certainly not when the government owned the banks, but often not when banks were privately owned either. As long as the customers of a bank have confidence in its ability to redeem its money certificates, however, they do not demand this of their bank. A few people test the system by redeeming their money certificates, but most people maintain confidence. However, once word has gotten out that someone was unable to redeem his money certificates, then everyone will be clamoring at the counters of the bank demanding commodity money.

15.2 The definitions of "fiduciary" and "credit" money

However great the confidence a bank's own community may have in it, it is doubtful that people in other communities, particularly bankers, share this confidence. While a bank is largely at the mercy of people in its own community should they be given reason to lose confidence, it has little to fear from foreigners, that is, clients of competing banks. If a bank defaults on claims held by foreigners, they have to overpower not only the bank but the community it is in. Most people prefer not to go to that extreme, so they avoid this problem by holding claims on foreign banks for only a few days. Thus, almost all intercommunity trade is in commodity money, though banks usually keep a small stock of foreign money certificates on hand, mostly for tourists. Rather than transporting commodity money in both directions, exporters hold foreign claims just long enough to deposit them at their own bank for money certificates. At the end of each week, their bank takes foreign money certificates to the banks of issue in exchange for domestic claims that their exporters have deposited, the difference being settled in commodity money. There is considerable variance from week to week and each bank must always have enough commodity money in its reserves to settle its differences at the clearing house, even if it expects the difference to reverse the next week. If it fails to do so, foreign banks will no longer trust it and will inform their clients not to accept checks written against that bank. Importers will have to carry commodity money to the other community to make their purchases. They can only acquire this commodity money by redeeming their money certificates at their own bank. Their bank, however, being temporarily out of commodity money, is unable to honor their money certificates and the disappointed importers will alert the community of the bank's illiquidity, causing it to be besieged by people presenting their money certificates for redemption. Even if the bank is solvent (the value of the claims it holds exceed the quantity of money certificates it has issued), people only want commodity money, and the bank cannot wait for even its shortest term claims to mature. Unless the banker can convince depositors to wait for some of these claims to mature, he will be forced out of business. This creates a strong incentive for a banker to keep his commodity money reserve above the week-to-week variance in the import-export trade. But, if money is issued only at the local level, it is not difficult to estimate the import-export trade and there is no necessity for legally-mandated reserve requirements.

Some people want to increase their money holdings and sell more than they buy, and others want to decrease their money holdings and buy more than they sell. This is what creates the variance in the import-export business as far as it reflects the general tendency to buy and sell. However, the Law of Large Numbers implies that, as the size of a community increases, the random variations in its import-export business tend toward zero relative to their quantity. Thus, there are two factors that determine the quantity of commodity money that a bank needs to keep in reserve relative to the money certificates that it has issued. The first is the size of a bank's clientele, which determines the degree of overall variance, and the second is how much of that variance is reflected in the import-export business. The latter is a function of the sophistication of a bank's clients (their tendency to deal with strangers) and the difficulty of overcoming transportation and tariff barriers between communities. Thus, banks in large, isolated communities have little to fear from variance in the import-export business threatening their reserves and can maintain small holdings of commodity money relative to the quantity of money certificates that they have issued. Banks with a small but sophisticated clientele, as in a big city with many competing banks, need to keep much larger holdings of commodity money. They can issue few money certificates for which they do not have a redemption fund of commodity money. When a shopkeeper in a city deposits his earnings in his bank at the end of the week, he is not surprised to find most of it written on other banks, whereas a farmer considers it unusual to accept money certificates issued by an unfamiliar bank.

If the clearing system is not fully developed, there may be occasions when commodity money is transported in both directions. Thus, the efficiency of the clearing system is a third factor in determining the quantity of commodity money that a bank needs to keep in reserve relative to the money certificates that it has issued. However, establishing an efficient clearing system is not a difficult task, so we will concentrate on the two most important factors, that is, the size of a bank's clientele and how much business they do with foreigners.

Money certificates issued in excess of commodity money reserves are called fiduciary money and are physically identical to the redeemable money certificates. If enough money certificates are brought in for redemption at once, not all of them can be redeemed and the excess is fiduciary money. Because a bank is liable for redeeming them immediately, unlike savings deposits which mature first, the issue of fiduciary money is not a credit transaction. What distinguishes a credit transaction is not that the creditor is trusted, for this characteristic is common to non-credit transactions such as the issue of money certificates, but rather that present phenomena is exchanged for future phenomena.

A common credit transaction is the acceptance of an IOU. People who have accepted an IOU but cannot wait for it to mature may exchange the receipt at a discount of its face value in the same way that they exchange money. As far as such IOUs circulate in place of money (they have a secondary market), they are called bills of debt and qualify as credit money. Thus, if p owes q a debt (as for capital whose value he cannot realize until he has assembled it with its complementary phenomena) and writes him an IOU, q may use it to settle a debt of his own with *r* or he may sell it to a bank, though in both cases at a discount of its face value, which can only be realized by holding the IOU until it matures. Thus, credit money is an addition to the stock of money (albeit a temporary one) because there are no money certificates devoted to backing them up. As Kindleberger dryly observes[18] "No bank balance is needed for a postdated check; the receiver hopes that there will be one when the due date rolls around." To function as credit money until then, the "pay to the order of" line must have been left blank and the hope for a bank balance must be widely held. If the issuer is well-known in the community and his reasons for postdating checks are clear (e.g. a farmer who waits to sell his crop), then his checks can function as money independent of any money certificates. The more transactions people cancel with debts owed them (the more they use credit money), the less their need for money certificates.

Besides private IOUs, certificates of deposit at banks as well as government and corporate bonds may function as bills of debt. Among bankers and capitalists, mortgages may also function as bills of debt though, in general, one thinks only of short-term credit instruments as credit money. What is important is that the issuer cashes the credit instruments on maturity regardless of who brings them in, which is prerequisite for their being passed from person to person as money in the meantime. Such bills should not be confused with savings deposits which pay interest but are first cashed before one can use them to cancel a debt. They do not augment the stock of money any more than liquid phenomena which one may consider selling for cash. For a phenomenon to qualify as money, it must itself be used to cancel debts, say, by writing a check or handing over a banknote. If it has to be sold to acquire money, it is not money itself or one would cancel a debt with it directly. This is true no matter how easy it is to sell a phenomenon. It is a serious mistake to define money as "very liquid phenomena," that is, very easy to exchange for money. Why exchange at all if very liquid phenomena *are* money?

Also, bills of debt should not be confused with credit cards, whose use has the opposite effect on the demand for money by reducing the frequency of payments and thus forcing people to keep large money holdings in anticipation of their monthly bills. The custom of paying rent or wages on a monthly rather than a weekly basis has the same effect. When one presents a credit card, one is not paying the retailer. The bank pays the retailer and it uses existing money, that is, it writes him a check. Signing a credit card slip does not create money. One is just promising to pay the bank later and, when one does, one will use existing money, that is, one will write them a check. The checking accounts, not the credit cards, are money.

The principle mistake of mainstream economics is addressing supply and demand rather than price and stock. However, assuming that all short-term credit instruments function as money is a close second. To require that bills of debt have an active secondary market and be readily accepted in payment of a debt is much more restrictive than merely requiring that they can conveniently be sold for money, which is the position of mainstream economics. The most important consequence of this mistake is discussed in section 16.2, but it is worth noting that defining money to include all short-term credit instruments also has the effect of soft-peddling an increase in the stock of money by a central bank. Lowering the interest rate through monetary expansion discourages savings so that, as the quantity of money certificates increases, the quantity of short-term credit instruments decreases. If money is defined as the sum of the two, the statistics conceal the increase in the stock of money. If one wants to quantify the issue of money, one should watch the growth of central bank assets. Central banks conceal their open market operations, however, and statistics on their assets may not be available even after the fact. Statistics on the stock of money are labeled "supply", which is confusing, and usually include a number of credit instruments, which is wrong, but are the only information released to the public (or to economists, who are actually more confused than the public). Thus, it is impossible to get an accurate measure of the stock of money. However, even without measurement, the proper definition of money is important, for serious policy mistakes can result from confusion here.

Commodity money and credit money are money in the narrower sense. Money in the broader sense is money certificates and money in the narrower sense that is not tied up in reserves against money certificates. The existence of fiduciary money implies no credit transaction and it can circulate indefinitely, whereas bills of debt are canceled on reaching maturity. They should not be confused with each other. An example of credit money is bearer bonds with a definite maturation date sold at less than face value and used to cancel debts while still current. Consols are always current (their maturation date is at infinity) and may also be used as credit money. However, promises by a bank (government) to redeem their money certificates, perhaps after an emergency is over, do not make fiduciary money into credit money. The value of money certificates that are no longer associated with commodity money is determined by speculation, as explained in section 10.1. As far as people believe such promises, they infer that the government does not intend to issue very many money certificates, so they do not add as great a price-change premium to their prices as they would otherwise. But fiduciary money is still very different from bills of debt which act as credit money. Bills of debt are always valued less than the money with which they are canceled. Their use as a medium of exchange causes their value to asymptotically approach the value of money, but not to exceed it. The value of purely fiduciary (fiat) money with no commodity money behind it proceeds from day to day without limits as discussed in section 10.1. How much the value of consols is affected by their use as a medium of exchange is also unlimited, which makes them behave very much like fiat money when used as a medium of exchange. A chart is given in figure 13 illustrating the several kinds of money.

Keep in mind that the concept of a central bank has not yet been introduced and, until it is, we are discussing a free-banking system, that is, one which uses a commodity money for its medium of exchange but where the clients of the several banks trade among themselves with certificates for commodity money. Each bank issues its own certificates for commodity money (in any quantity they want to) and people do not hold the money certificates of other banks for fear that they will not be redeemed. In addition to money certificates, bills of debt also circulate as a medium of exchange until maturity, as does commodity money that is not held in reserve against money certificates. Another chart will be given in section 16.3 which includes the existence of a central bank.

Figure 13

Types of Money in a Free Banking System

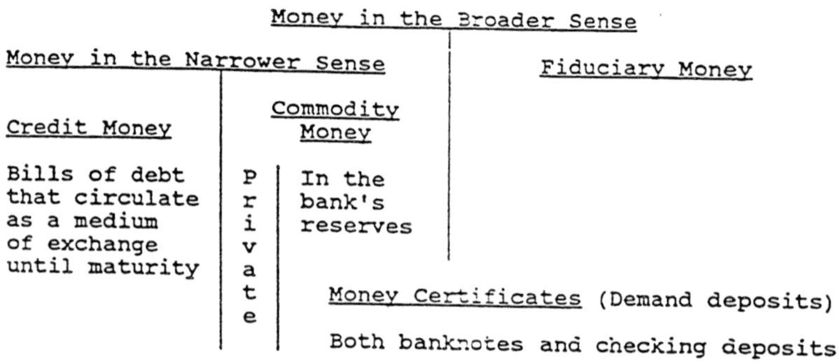

This chart (figure 13) is somewhat misleading in that the reserves devoted to the redemption of money certificates appear to be composed entirely of commodity money. While this is often the case, it must be remembered that the purpose of a reserve is to satisfy the needs of importers whose suppliers, or rather whose suppliers' bankers, do not wish to hold domestic money certificates. It is not at all to satisfy customers who have lost confidence in a bank. This latter event occurs catastrophically and it can only be prevented, not halted. The needs of importers can sometimes be satisfied with foreign bills of debt, however, allowing a banker to invest part of his commodity money, which otherwise would not be earning interest. Foreign bills of debt in a banker's portfolio should not be mistaken for foreign money certificates, which a bank never holds past the weekly clearinghouse settlement, as they bear no interest and can be redeemed immediately for commodity money. Theoretically, these bills of debt are no different than commodity money because, being foreign, they can only be purchased with commodity money and it is with commodity money that they are canceled. Practically, however, the habit of small banks to hold claims against larger banks tends to concentrate commodity money in the reserves of a few large banks. Speculation on the value of money is made more difficult because of arbitrary changes in policy on the part of these large banks, which may weaken support for a commodity money standard. If people become accustomed to using only money certificates and not commodity money in their transactions, the disassociation of money certificates from commodity money is facilitated, the effects of which are the subject of much of the rest of this chapter.

15.3 The limit on the issue of money certificates

There is an incentive for a bank to issue as much money as possible, for it can be used to buy capital away from people who have to pay for it with actual savings. The additional money diffuses out into the economy in concentric circles around a bank, inflating people's money holdings, which causes them to buy more than they sell in order to change their excess liquid phenomena into more durable wealth. According to the Law of Price Adjustment, this excess of buying over selling tends to raise the price of phenomena. Thus, people farther removed from a bank are being taken advantage of when they sell at the low prices that they are used to, not realizing that the prices for their phenomena will rise. When understanding among those farther removed from a bank increases, they add a price-change premium to their prices; that is, they raise them early in anticipation of having to raise them later. Until a community actually rejects its money certificates, however, there are people farther removed from the bank who refuse to raise their prices, causing their price-change premium to fail to keep up with the inflated money holdings of those close to the bank. Thus, people close to an issue of new money continue to profit from the devaluation of their community's money certificates until very close to the end. This effect is aggravated by the habit of those in ignorance of the situation to blame the problem on speculators who add a price-change premium before the general rise in prices.

While we are primarily interested in asymmetrical understanding of a depreciation in money, profit may also be made by predicting an increase in the value of money, the most important instance being of commodity money when banks fail. Another instance is the attempt to revalue money certificates at their previous level after they have been devalued to pay for an expense that a government did not wish to neglect but for which it could not convince people to pay taxes, such as an aggressive war. It is sufficient to note that different people are in a position to speculate on increases in the value of money than those who were hurt by the devaluation of the previous monetary unit. Temporarily abandoned commodity-money standards are sometimes completely abandoned because economists assert that returning to a commodity-money standard requires revaluing money certificates at their previous level, which is not politically feasible. In fact, it is sufficient to redefine devalued money certificates at their current level. Historically, opposition to a commodity-money standard is considered a populist cause, but these movements have always occurred after a commodity-money standard was temporarily abandoned to pay for a war, after which economists insisted that the currency be revalued at its pre-war level. This imposed hardships on people such as farmers without money-denominated assets. However advisable it may be to wage a war that people will not pay for with taxes, once done, there is no necessity for deflation. Temporarily abandoning a commodity-money standard must have costs lest it be done too frequently, but making farmers bear the costs has incited populist movements that eventually proved fatal to commodity money while too-frequent abandonments did not. Hereafter, however, we will assume that all speculation is on the absolute devaluation of money certificates (though not necessarily relative to foreign money certificates), a procedure whose relevance can be verified by recent history.

By the time an old issue of money diffuses out to people farther removed from a bank, those closer to the bank have a new batch. As long as those farther from the bank continue to buy in the community, they have their wealth siphoned off. This gives them a strong incentive to buy out of town where prices are still low. Out-of-town sellers naturally encourage this, though they demand payment in commodity money as they have no use for foreign (to them) money certificates. Exports having dropped off because of high prices in the inflationary community and imports having risen because those same high prices are driving domestic consumers to foreign markets, there is an unfavorable balance of trade. This exodus of commodity money is a simple application of the theory of marginal utility, for money, like any phenomenon, is put to its most useful purpose. If individual units of commodity money can buy very little in a community pursuing an inflationary policy and more in neighboring communities, they go to those neighboring communities. Or, put another way, a community uses the cheapest phenomenon available that satisfies a particular need, in this case the need for a medium of exchange. When importer's demand for commodity money rises to meet the declining reserves (because exporters are no longer bringing in foreign commodity money), a bank has to halt its issue of money for fear of a bank run prompted by an importer's failure to redeem his money certificates. Thus, it is doubtful that a banker will press his issue of money beyond its natural limit.

Even if bankers do not directly purchase capital because it creates class resentment between the favorites of a bank, who blame the less favored for a lack of patriotism when they shop out of town, and the less favored, who blame those close to the bank for their privileges, the same upper bound is set on the issue of money. If a bank favors everyone by lowering its interest rate (assuming that the bank does not discriminate when granting credit to favor particular programs, in which case the stratification of society is the same as if it had directly bought into those programs), the low rate of interest attracts out-of-town borrowers who demand their loans in commodity money. These borrowers profit through arbitrage by reloaning the commodity money in their own community at the higher interest rates prevalent there. To prevent this practice, whenever banks believe that a demand for commodity money is for foreign loan markets, they charge a premium on the redemption of their money certificates; that is, they give less commodity money than the face value of their money certificates indicate that they are worth. This may be accomplished by charging fees, by giving out worn or damaged commodity money, or by moving the redemption center away from the ports. As much as this policy discourages the export of commodity money, however, it also discourages its importation in the event that foreign interest rates drop below the domestic rate. Foreigners, who might then redeem their own money certificates for commodity money to be imported into the community charging the premium, abstain for fear that they could not repatriate their commodity money should the relation reverse and the community charging the premium began devaluing its money certificates more than their own. Thus, this policy only works for creditor communities (ones saturated in capital), as their rent is naturally lower than foreign rent. As interest is one component of rent, domestic prices (the other component of rent) have to be very low relative to foreign prices to cause domestic interest rates to be higher than foreign interest rates. Debtor communities which are dependent on the importation of capital only harm their own interests by following this policy. Whether it works or not, it is largely counteracted by the policy of foreign banks to pay bonuses to importers for commodity money deposited at their banks. Thus, if every bank pays as much extra to the importers of commodity money as it short-changes the exporters of commodity money, they cannot prevent arbitrage.

While the rate of saving is determined by the relation of the stock of capital to saturation, there may be obstacles which prevent direct investment of labor in capital production, especially across national borders. It was to circumvent such obstacles that the institution of banking was developed; if they betray that function by excessively issuing money, they lose savings, especially from foreigners. There are many people with a large portion of their wealth in short-term credit instruments who do not consider directly investing their labor in capital production even in locations where the stock of capital is much less than saturation. They may fear confiscation, they may not want to leave their homes, or, most likely, they may already have jobs somewhere and cannot afford to start their own business. They feel no obligation to invest at home, however, and invest wherever they get the highest return. They abstain from buying long-term credit instruments because of their fear of monetary devaluation in the community that they patronize. In fact, unexpected devaluation of the money in which so many of their assets are denominated is a constant specter for such people and they are prepared to shift their savings at the slightest drop in the interest rate. While the quantity of short-term credit instruments bought in the world is uniquely a function of wealth, what is important to bankers is the savings in their community. If they lower their interest rate by issuing money, they lose savings as the wealth in short-term credit instruments is redistributed around the world.

The intercity (or international) distribution of commodity money is a function of the distribution of the demand for money holdings. As individual variations among people in their need for liquidity tend to cancel each other out according to the Law of Large Numbers, there is no necessity for large movements of commodity money between communities. (Large relative to the overall demand for commodity money; absolute quantities shipped between communities rise with larger numbers of people but their ratio with overall demand converges to zero.) A community experiences a loss of commodity money only if everyone is encouraged to buy more than they sell because their money holdings are inflated with fiduciary money. This has nothing to do with whether a community is rich in capital or not; debtor communities export commodity money when they buy capital and import commodity money when they sell finished phenomena. Creditor communities do the opposite. It is not the quantity of wealth in a community that determines its balance of trade but whether the proportion of their wealth that is kept in money holdings is increasing or decreasing.

If everything that the people of a poor community need must be imported, they can only get it by exporting what people in other communities need. If they have nothing to export, they do not import anything either and their needs remain unsatisfied though they have a balance of trade. That their needs are unsatisfied is a result of their poverty and not their trade policy. If anything, they should avoid devaluing their money certificates so the purchase of credit instruments denominated in the domestic currency is not discouraged. I leave it to anthropologists to judge the effects of foreign-owned capital projects in undeveloped countries. If the indigenous people feel that it is helping them join civilization, they should not drive it away with threats of confiscation. If they feel that they are being unjustly forced into working for foreigners, they should take up whatever form of production suits them. But, being poor, they should realize that an expansion of any form must be financed by the sale of credit instruments to foreigners rather than to natives and that foreigners are not going to buy credit instruments denominated in a depreciating currency.

However romantic a foraging economy may seem to anthropologists, that is no longer an option. Undeveloped countries are either going to work on foreign-owned capital projects or they are going to buy the capital to operate such projects (and the weapons to defend them) by selling credit instruments to foreigners. Industrial countries usually have the advantage both in war and in the operation of large capital projects, but they may not have the support of their people when it comes to fighting over foreign resources. If the government of an undeveloped country is humane and fiscally responsible, it may acquire enough investment and moral support from people in industrial countries to allow it to operate its own capital projects. But most governments of undeveloped countries are just playing the several industrial nations against one another and have no aspiration for independence. No particular group has a moral imperative to own the natural resources in an area and one can always expect to find fighting where resources are concentrated. Somebody must own them; it is their nature not to be accessible to a cottage industry. The Creator made a prize of natural resources by concentrating them geographically and with the same stroke made them accessible only to people backed up with lots of capital and weapons. This may or may not be the natives of the region.

15.4 The possibility of alliances among banks

Whether banks directly purchase capital or loan it at interest, the limit on a bank's issue of money is a function of its size and isolation from competitive banks. Small, competitive banks in large cities have little opportunity for excessive issues of fiduciary money. This is because the customers of one bank are always in contact with customers of other banks from whom they can buy phenomena but who will not hold foreign (to them) money certificates. Also, borrowers and savers habitually shop around for the lowest or highest interest rate available, respectively. In such communities, even the smallest quantity of fiduciary money is quickly brought back to its bank of issue for redemption in commodity money, which is then deposited at competing banks. This limit on the issue of money is severe; even large central banks with the ability to raise armies feel their monopolies shaken by innovations in data processing and communications which allow people more contact with competing financial institutions.

Banks with such worldly clients no doubt envy the protection which isolation provides for their colleagues in the country. But to leave their small share of the bustling market of the city and establish a monopoly in the country to profit from devaluing the currency there is not advantageous. Thus, worldly banks might consider an alliance among themselves as a means to simulate the protection that isolated banks enjoy without sacrificing the brisk trade of the city. The possibility of an alliance, however, is most attractive to banks that have devalued their currency with such recklessness as to deplete their reserves to a dangerous level. In fact, more prudent banks are quite happy to see their competitors fail, for nobody gets their commodity money except the sounder banks when the weaker ones lose their reserves. The demand for an alliance is, then, nothing more than weak banks asking stronger ones to help them out of a tight spot. Having nothing to offer in exchange for such help, their reserves and their clients going to the stronger banks when they fail, it is unlikely that they will receive much sympathy from their competitors. A strong bank receives no profit, only loses, from allying itself with a weaker bank. Instead, inflationists are cheered on to their demise by their competitors and all thoughts of an alliance go down with them.

Section 16

16.1 The establishment of a central bank

Most people peacefully accept a banker's refusal to finance their project on the grounds that, it being his business, he probably knows best which projects will yield a return on the investment necessary to capitalize them and which are submarginal. Some people, however, find ideological reasons to convince themselves that their ventures are worthwhile despite their apparent foolishness. Ideologists being difficult to sway, one can expect an effort to be made to lower the rate of interest to include such ventures or to buy capital directly with newly-issued money certificates. All the banks in as large a geographical area as those undertaking the venture can control (usually a nation) are organized in a central banking system which confiscates all the reserves. No competitive bank can be left because, if it does not devalue its currency to the same extent, it will siphon off all the commodity money of the central bank. Furthermore, people in that geographical area, now clients of the central bank, are not allowed to buy phenomena from foreigners for, if any of their money certificates were to escape to competitive banks, those banks would bring them back for redemption in commodity money. Though a central bank can refuse to give up its commodity money reserves when its money certificates are brought back for redemption, pressure from other nations may prevent this while none of a central bank's customers have the ability to resist a tariff. Exports are also hurt by the price-change premium which domestic sellers add to their prices, so very little phenomena is sold to foreigners unless they are forced to accept it. With these measures, the clients of a central bank are effectively isolated from any competitive bank's customers and they no longer need to redeem their money certificates for the purpose of importing foreign phenomena (though neither are exporters bringing in foreign commodity money), so the central bank's reserves are stable.

Though it may seem to take a considerable amount of effort to monopolize the issue of money certificates, a central bank is supported by alliances of suppliers when establishing tariffs and other isolationist measures. It may also convince people that small competitive banks tend toward failure for, as there were only small competitive banks before the central bank confiscated their reserves, any that failed can be used as examples. This latter technique is only effective if the laws had been lenient on such failures. Thus, nationalization of the right to issue money certificates is usually preceded by laws which encourage private banks to issue excessive amounts of money certificates. Among other things, such laws provide tax money to bail out insolvent banks. Since history does not provide one with any examples of a free-banking system, the argument about whether bank failures occur because of, or in spite of, government regulations and bailout programs is unwinable unless brought into the clear light of abstract reasoning.

The term "regulation" is meaningless without reference to the basic framework in which banks operate. A stable system can be governed by the usual laws against criminality that apply to all businesses, while an unstable system requires a vast regulatory bureaucracy and is still plagued with corruption. It is naive for people who dislike big government to advocate deregulation in the latter case, but it is also wrong to assume that the existence of a central bank is part of the regulations which attempt to prevent corruption. Central banks and regulatory bureaucracies are associated with one another, not because they both oppose an inherent instability in banking, but because the existence of a central bank creates an unstable system that requires constant policing.

Criticism of free banking invariably involves attacking a straw man. It is assumed that everybody is agreed on the need for and existence of tax money to bail out insolvent banks and that "free banking" means freedom to dip into that fund without censure. The claim that free banking produces sound money is then ridiculed as equivalent to the claim that banks would compete to see which of them can make the more conservative loans. The straw man is easily slain by observing that, in such a situation, they would actually compete to see who can make the more risky loans. For example, Kindleberger criticizes free banking advocates in this way[19]:

> *They assert their readiness to allow any bank, company, or person to issue "money" and try to gain its acceptance; the market, they claim, will decide who issues good money. Instead of worrying that "bad money" will drive "good money" into hoarding and export, in line with Gresham's Law, they maintain that good money will drive out bad, as issuers vie for acceptance.*

In fact, the stability of a free banking system does not depend on the acceptance of a bank's money by foreigners but, just the opposite, it depends on their mutual distrust and haste to redeem one another's money certificates. By the time an old issue of money diffuses out to people farther removed from a bank, those closer to the bank have a new batch. As long as those farther from the bank continue to buy in the community, they have their wealth siphoned off. This gives them a strong incentive to buy out of town where prices are still low. Out-of-town sellers naturally encourage this, though they demand payment in commodity money as they have no use for foreign (to them) money certificates. Exports having dropped off because of high prices in the inflationary community and imports having risen because those same high prices are driving domestic consumers to foreign markets, there is an unfavorable balance of trade. This exodus of commodity money is a simple application of the theory of marginal utility, for money, like any phenomenon, is put to its most useful purpose. If individual units of commodity money can buy very little in a community pursuing an inflationary policy and more in neighboring communities, they go to those neighboring communities. Or, put another way, a community uses the cheapest phenomenon available that satisfies a particular need, in this case the need for a medium of exchange. When importer's demand for commodity money rises to meet the declining reserves (because exporters are no longer bringing in foreign commodity money), a bank has to halt its issue of money for fear of a bank run prompted by an importer's failure to redeem his money certificates.

Thus, it is doubtful that a banker will press his issue of money beyond its natural limit. By prudence he can hope to survive, but by no policy can he hope for the acceptance of his money certificates by foreigners, who will always be quick to redeem them. Bankers do not "vie for acceptance" because that is unattainable and, fortunately, the system does not depend on it.

16.2 The effects of canceling credit money with fiduciary money

The use of credit money (bills of debt) is an alternative to using money certificates and thus a negative influence on their value. It is sometimes thought that, if the issue of money certificates, which is also a negative influence because it increases their quantity, is used only to eliminate the first negative influence, the value of money remains stable. Thus, the argument continues, if a central bank refrains from issuing money certificates except to buy bills of debt which are then held until they mature, the quantity of fiduciary money in circulation is made directly proportional to the quantity of credit money removed from circulation, leaving the stock of money in the broader sense unchanged. This would work if it were not for the fact that the quantity of credit instruments removed from circulation (sold to a bank) is itself a function of the quantity of fiduciary money issued by the bank. When a person writes an IOU for phenomena received, it has a face value greater than the present price of that phenomenon or the seller would demand immediate payment. If the seller of that phenomenon is unable to wait for the IOU to mature, however, he can sell it to a bank at a discount of its face value, thus receiving only what he would if he demanded immediate payment for his phenomena. Because of his weakness and the bank's relative strength when it comes to the ability to wait for an investment to be realized, the bank collects interest on that IOU. This strength of a bank's, like anyone else's, is a function of its wealth. However, because a central bank has the ability to arbitrarily increase its wealth with fiduciary money, it has a disproportionate advantage. This allows it to buy credit instruments at very close to their face value even when they still have some time to mature. This practice greatly increases the demand for credit instruments instead of immediate payment because people know that they can sell them to a bank. Thus, credit instruments are sold to a bank that were never really a part of credit money as they were never exchanged from person to person, which would have decreased the need for money certificates, but were sold to the bank immediately upon being written. Existence alone is not a sufficient condition for credit instruments to be included in the stock of money; there must be an active secondary market for them in the community.

The so-called "elastic money market" argument that restricting the issue of fiduciary money to the purchase of short-term credit instruments only changes the composition of money (from clumsy bills of debt to more convenient money certificates), but not its value, is academic in a free-banking system and wrong in the presence of a central bank.

1) In a free-banking system, the several banks, each issuing their own money certificates, are limited by the fear that their clients may buy phenomena from the clients of competing banks and that those people, through their own banks, will bring the foreign (to them) money certificates back for redemption in commodity money. If a bank lowers its interest rate by buying credit instruments at very close to their face value (there is no other way to buy credit instruments in quantity), people indebted to their own banks refinance their debts by selling credit instruments (for commodity money) to the bank that is expanding its issue of fiduciary money.

2) In the presence of a central bank, there is no limit to the quantity of money certificates that a central bank can issue, though the more credit instruments that they buy, the less profit that they make by holding them. This is because, as they buy larger and larger quantities of credit instruments, the purchasing price asymptotically approaches their face value even though the face value can only be realized after some time.

Thus, the point is whether or not banks have been monopolized by a central bank; the composition of their assets is a matter of relative indifference.

Commercial banks sometimes purchase commercial bills of debt, but central bank assets are almost entirely in government bonds which are either bought directly from the government or from private dealers who never intended to hold them to maturity. In the former case they are called "new", and in the latter case they are called "old." There are several reasons why a central bank prefers to purchase old bonds. First, if it purchases bonds directly from the government, it is difficult to claim that the purchase removes credit money from circulation. Many people are not aware that private bond dealers purchase government bonds for the sole purpose of selling them to the central bank. Because these dealers are private, it appears that the bonds are removed from an active secondary market. Also because these dealers are private, their purchases appear to be of patriots answering the call to buy government bonds. This tends to sanctify government expenditures which people did not vote to support with their taxes. Most people understand that the expenditures that exceed taxes are not wanted by the citizenry or they would vote for a tax to support them, but people can sometimes be deceived into thinking that "government debt" is held by patriots rather than by the government itself (the central bank) or by commercial banks, which are effectively a part of the government. Finally, by purchasing old bonds, a central bank can placate the opponents of deficit spending by pointing out that old bonds have already contributed to the deficit and their movement through the secondary market does not change that. This is a weak argument because such bonds were only purchased in the first place on the assumption that the central bank would buy them, and now it is making good on that promise.

The tendency of a bank to limit its issue of money to the granting of short-term credit is also influenced by the thought that claims that mature quickly are better reserves than long-term claims. This is because, if people's confidence in their money certificates is lost, a bank only has a short time to wait for such credit instruments to mature and make the bank liquid again. Confidence in a bank does not fail gradually, however, and in the event of a bank run, a bank can no more wait for its short-term credit instruments to mature than it can wait for its mortgages to mature. The tendency to prefer short-term loans over long-term loans is best explained by a bank's fear of adding too small a price-change premium to its interest rate on long-term loans because it cannot anticipate how far its own issue of money will go in the more distant future. This is more the case of small banks who do not themselves participate in the decisions about issuing money. Commercial banks may fail if they enter long-term agreements at a fixed interest rate with their clients, only to have the central bank press the issue of money to such a point that the currency paid by debtors to the commercial banks in accordance with their long-term agreements is almost worthless. Thus, as long as commercial banks are not kept informed about future policy changes concerning the devaluation of the currency, they are hesitant to make long-term commitments. Legal tender laws prevent banks from making loans payable in more stable phenomena, such as finished phenomena produced with the capital bought with a loan.

No bank that undertakes the issue of fiduciary money can hope to remain liquid in the sense that businesses are required to be; that is, to meet all of the obligations that might be demanded of it on any particular day. Money certificates (demand deposits) can be enforced against a bank at any time without warning and fiduciary money is, by definition, money certificates issued without cover. Such banks can only hope for solvency, that is, the ability to redeem their money certificates after selling their assets. Being forced to sell assets, however, is not a position that banks want to get into, so no owner purposefully makes loans so recklessly that he gets to that point. Members of a central bank system might, though, because they do not actually have to sell their assets to gain liquidity but need only put them up as collateral for an emergency loan (an advance) from the central bank or sell them temporarily to the central bank at a discount. Owners of such banks can be more liberal in making loans than if they were independent because they need never fear seeing their property get auctioned off. They must only feel the embarrassment of having to ask the directors of the central bank to help them out. Such loans come with strings attached, however, which has caused commercial banks to loan money to one another to avoid having to ask the central bank for credit. For the purpose of collateral, long-term claims such as mortgages serve better than short-term credit instruments.

Even if they have been so reckless that it is obvious that they cannot pay off a loan from anyone, most governments have "insurance" programs to bail out insolvent banks with tax money. The incentives to issue money are so great, however, that insolvency among banks quickly outruns a government's ability to tax its subjects and most of these programs are operating on faith alone. This can actually go on for a long period of time. People are gullible.

Insurance or bailout programs do not require the existence of a central bank. In fact, the first step toward monopolizing the issue of money certificates is the establishment of such programs. By making banks indebted to the government, they become semi-public institutions and it is not a big step from there to establish a central bank. More importantly, the availability of tax money to bail out insolvent banks causes them to issue an excessive amount of fiduciary money and, therefore, to eventually fail. This is bad publicity for the institution of private banks and it makes it easy to convince people that a central bank would solve the problem of bank runs. There is no real difference between an insurance program and a bailout program, that is, between giving tax money to people whose money certificates have been defaulted on or giving it to banks so that they do not default on their money certificates. However, because of the publicity aspect, before the institution of a central bank it is common to bail banks out subtly so that they still appear to be private institutions when they fail. After the institution of a central bank, insurance programs are more common.

16.3 The operation of a central bank system

A central bank acts as a "banker's bank" and thus adds another level to the chart of the several kinds of money. See figure 14. There is now a difference between banknotes and checking accounts at the level where they were previously synonymous, that is, the level once occupied by independent banks and now monopolized by a central bank. That difference is that banknotes may be held by private parties, whereas

Figure 14

Types of Money in a Central Bank System

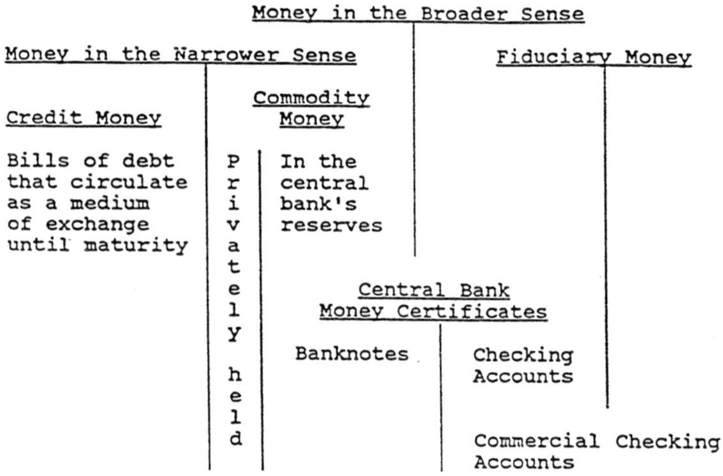

only commercial banks may have checking accounts at the central bank. Furthermore, commercial banks are no longer allowed to issue their own banknotes but only to open checking accounts for their clients. Thus, while people still have the option of holding commodity money, banknotes, or checking accounts, these are now the distinct representatives of the primary, secondary and tertiary levels of the monetary system, respectively. Where legal tender laws prohibit the enforcement of contracts written in terms of commodity money and the central bank has given up redeeming its money certificates in commodity money except to the most powerful claimants (usually only other nations), the primary level tends to fade into the background. This creates a fiat system in which checking accounts are redeemable in banknotes and banknotes are valued only because of the government's promise not to issue too many. To the extent that a government allows its banknotes to escape into foreign countries, however, too great an issue of banknotes causes it to fear foreigners forcing redemption in commodity money. Also, international law may fix exchange rates, the importance of these laws to a country depending on its relative size and military strength.

Commercial bank accounts exceed their deposits at the central bank by a multiple fixed by law or by custom. Thus, banks are interested in increasing their deposits at the central bank either at the expense of private banknotes or by grant. Private banknotes are only deposited at commercial banks if people trust banks and checks are widely accepted. People's trust can be engendered with propaganda and their acceptance of checks by a vigorous pursuit and punishment of check bouncers. Laws which sanction the confiscation of people's checking accounts are weakened so that people who fear the government need not keep all their money holdings in banknotes. People are encouraged to lengthen the period in which they pay ongoing expenses (rent, wages, credit cards, etc.) because small, frequent payments are made out of petty cash whereas large infrequent payments are made by check for security and tax purposes. Infrequent payments cause people to deposit their banknotes in checking accounts, allowing the bank to deposit them in their central bank account as reserves against multiple checking accounts. This increases the quantity of fiduciary money and hence decreases the value of money. Infrequent payments also cause people to keep larger money holdings which in turn increases their demand for and hence the value of money. The former is usually the stronger effect.

A central bank can increase commercial bank reserves by fiat, that is, by issuing as many banknotes as are needed to satisfy customers who wish to close their accounts at commercial banks. Politically, however, this method can only be justified in emergencies. For financing day-to-day expenses, the treasury sells bonds either to the central bank or to commercial banks. Commercial banks' accounts at the central bank are only barely sufficient for the checking accounts multiplied on them, however, so they do not have any extra funds to buy government bonds. The central bank, however, can buy government bonds with its own money certificates (created *ad libitum*) which are then deposited at commercial banks. These banks deposit them in their central bank accounts as a reserve against multiple checking accounts of their own, which can be used either to buy government bonds or to loan at interest to private parties. The whole routine, of course, is a subterfuge for economists who believe that canceling credit money with fiduciary money does not increase the stock of money. See section 16.2. In fact, the stock of money increases more than it does under the primitive method where a government simply monopolizes banks and then issues money certificates *ad libitum*, but without a tertiary level. The method of selling bonds has another effect besides inflation, though, because the treasury has to pay them back at interest when they mature. Since nothing a government does can be called an investment in the sense that it brings a return, future administrations monetize this debt, which is why there is so much political pressure to continue issuing money.

A central bank is a monopoly: It enjoys the exclusive right to issue banknotes and, while commercial banks may open checking accounts, they are subservient to the central bank because their checking accounts may not exceed a fixed multiple of their deposits at the central bank. Commercial banks are competitive with one another, however, which pacifies anti-monopolists and casts doubt on the assertion that free competition is sufficient to limit the issue of fiduciary money. Why does competition between commercial banks not restrict the issue of fiduciary money? It does prevent any particular commercial bank from increasing its checking accounts by the full multiple over central bank money certificates it receives and which it wishes to use as reserves against multiple checking accounts.

Let the allowed multiple of checking accounts over central-bank money certificates held by commercial banks be $m > 1$. If c central-bank money certificates are deposited at a particular commercial bank, that bank cannot loan cm in checking accounts to its customers, for they will buy phenomena from clients of another bank which will demand that the first one pay it in central-bank money certificates of which it has only c and not cm. Instead of loaning cm, however, a bank loans $c\left(1 - \frac{1}{m}\right)$ so that, when this amount is brought back for redemption by a competing bank, the first bank still has $\frac{c}{m}$ central-bank money certificates. This is exactly what it needs to cover the checking account it gave to whoever initially deposited the *c* central-bank money certificates. Thus, the effect of this initial deposit moves from bank to bank, each one increasing the stock of money by $1 - \frac{1}{m}$ of the amount that the previous bank increased the stock of money. The total increase is then $\sum_{n=0}^{\infty} c\left(1 - \frac{1}{m}\right)^n$. Letting $r = 1 - \frac{1}{m}$ and recalling the identity $(1-r)(1+r+\ldots+r^{k-1}) = 1-r^k$, the summation out to $n = k-1$ is $\frac{c(1-r^k)}{1-r}$. But r^k goes to zero as k approaches infinity because $m > 1$ implies that $0 < r < 1$, so the total increase in checking accounts after diffusion is $\frac{c}{1-r} = cm$; exactly what one expects it to be. The same process takes place in the absence of a central bank with the exception that, with a physically limited commodity money being the only accepted reserve, no bank can gain reserves except at the expense of another. A central bank can arbitrarily increase a particular commercial bank's reserves by buying assets from one of its clients or by fiat.

A large government spends a tremendous amount of money and, for good or evil, it is important to understand where they get that money. We are now at a point where we can make such an analysis. Whenever examples of government waste appear in the news-media, the expenditures are always referred as coming from "tax money." Some of them do, though taxing citizens lacks subtlety and collecting fines can be dangerous. With the tendency toward conglomeration, however, the importation of controlled substances such as narcotics is usually just one facade of a multinational and multinationals rarely risk their whole business by fighting. Corporations are usually fairly placid and just pay whatever fines or bribes are necessary, either to establish a puppet government in a small country or to keep the existing one in a large country happy. Civilians are also fairly placid, especially in countries where citizens are encouraged to buy houses and automobiles on credit. Governments that build railways and apartments rather than freeways and suburbs have less control over their subjects. A workingman who spends three-fourths of his income on house and car payments is poor and stable while one who spends an equal amount on rent and public transportation is just poor. Also, traffic fines and retail narcotics fines are the major source of income for local police, who must be kept happy lest they form alliances and fight with the national army. (It should be noted that the construction of freeways rather than railways is motivated by military as well as social considerations. However, the construction of suburbs rather than apartments is motivated entirely by the desire to bind people to property that can easily be taken away from them should they be bad. This is evidenced by the fact that laws which establish central banks are referred to as housing acts.) Nevertheless, such mundane measures would not be sufficient to assure stability if a large government had to pay for its expenditures with taxes and fines alone. Governments need a more subtle source of income, and one that is unavailable to gangs and local police.

In fact, the great majority of government expenditures are paid for with government debt. It is natural to inquire about who they owe it to and whether their creditors are in a position to influence the government. If their creditors were in such a position, the system would be less stable than direct confiscation. The system is stable, however, and for the simple reason that they owe the debt to themselves. While anyone can buy government bonds, they are usually not competitive with private bonds. When they are, they crowd out that much private investment. That situation is unusual, however, because when civilians, especially foreigners, buy bonds, the treasury has to pay interest to people that it cannot easily get the money back from. Far more common is when the treasury sells bonds to the central bank (sometimes through intermediaries), the central bank writes the treasury a check which they deposit in a commercial bank, and commercial banks use that check as a reserve against deposits of their own totaling several times more than their reserve (*m* times more, with *m* defined in the preceding paragraphs.) Since the central bank and the treasury are both branches of the government, they have effectively just printed some money and spent it. The word "debt" is used only to obscure the process and does not have any meaning in this context. The amount of money that a government can print (money is not physically printed but is just recorded in checking account balances) is limited only by political pressure to slow inflation. This is a weak limit because the price level is a virtually meaningless statistic. It is a limit, however, because the government's purpose is not just to make money but to buy things for themselves with it and, if people abandon the currency, the government cannot buy anything.

The term “national debt” is probably the most misunderstood concept of economics. The national debt is the sum of all outstanding government bonds. If the national debt increases over the course of a year, then the government has run a deficit. If the national debt decreases, they have run a surplus. Government bonds are sold mostly to the central bank, some to commercial banks, and almost none to private savers. One possible exception is countries with a large trade deficit. This implies that foreigners own some of their assets and, while they usually prefer real estate and businesses, they may take government bonds. A country can have a trade deficit (it imports more than it exports) only by selling stocks and bonds to foreigners. In this way, a nation of a billion people is no different than a single farmer. A farmer raises some produce for his own family’s consumption and some to be sold on the market. By necessity or choice, a farm may be mostly self-sufficient or it may concentrate on cash crops. A farmer can buy more than he sells only by borrowing money or by selling equity. If he sells all of his equity then he becomes a tenant and, if he sells some of it (like a tractor), then he must rent equipment that he once owned. There are two reasons for him to have a trade deficit. One is that he had a poor harvest and he needs money to buy food and provisions until he has a better one. The other is that he wishes to improve his farm but his personal savings are not sufficient to pay for the needed equipment. As his fortunes have varied, he is left simultaneously paying off debts and receiving income from notes that he holds against other farmers, but overall he is either a net debtor or a net creditor. Similarly, nations are either net debtors or net creditors. This should not be confused with the term "national debt", which refers only to government bonds.

The above analogy is a good one and it often helps to think of each nation as a farm. Some countries have trade deficits because they are temporarily unable to meet their needs. This is not the same thing as being poor (poor countries do not have trade deficits because nobody gives them credit) but is a matter of having an uneven income stream. The unevenness is either due to vagaries in their output or because they are growing. Other countries are well-provisioned but consume all that they produce and have trade deficits because they must turn to foreigners for investment. That they have a low savings rate is sometimes cultural (they are decadent), but is usually caused by a large national debt. Also, money made in the underground economy cannot be invested because, if one purchases publicly traded stocks and bonds, the tax authorities will take both the principle and the interest. Without any legal recourse if a debt goes bad, few people in the underground economy have the ability or the desire to be non-bank lenders (loan sharks) and so they spend their money frivolously on consumer goods. Thus, oppressive tax rates directly cause low savings rates.

Whatever the cause of their trade deficit, it is financed by selling either stocks or bonds to foreigners. In the former case, foreigners will either have a portfolio of many minority stock positions or they will buy real estate and businesses outright. In the latter case, foreigners will buy either corporate or government bonds. Direct investment, rather than portfolio investment, hurts the people's pride because they must work for foreigners. The purchase of government bonds, rather than corporate bonds, hurts the central bank's pride because it is no longer completely free to control the value of its money certificates.

Government bonds which offer a high enough interest rate to compete with corporate bonds crowd out that much private investment. However, this situation is rare, as government bonds are usually sold only to the central bank, though possibly through intermediaries, to conceal the creation of new money from the public. This is why I said that government bonds are sold mostly to the central bank, some to commercial banks, and almost none to private savers. However, even in this case, where government bonds offer a lower interest rate than corporate bonds, foreigners often prefer them over corporate bonds. This is because foreigners have unreliable information about the trustworthiness of private bond issuers and no convenient means of pursuing bad debts. Thus, the few government bonds in the hands of private savers are usually in the hands of foreign rather than domestic savers.

One exception to this is during an emergency, like a war, when people buy government bonds regardless of the interest rate, partly out of patriotism and partly because rationing leaves them with extra money. Another exception is, ironically, when a stock market bubble is about to burst. Investors know that the central bank will continue to roll over the government's debt, that is, current government bonds will be paid off by selling new ones to the central bank. Since no one is going to roll over corporate debt after the crash, government bonds are viewed as a safe place to put one's money when the stock market becomes volatile and uncertain. I say that this is ironic because the government (specifically, the central bank) caused the bubble in the stock market. Nevertheless, they can later paint a picture of a "flawed" capitalist system prone to irrational exuberance and contrast it with the government, which is presented as a model of creditworthiness that can easily sell bonds, even in bad times.

16.4 Central banking in import-dependent countries

If a central bank has defaulted on its money certificates only with regard to its own people but still redeems them in commodity money when approached by another country, the same effect that appears among community banks takes place at the international level. The central bank of a small country with a brisk import-export business is more limited than one of a large, isolated country because the small country fears a loss of its commodity money reserves. But what if central-bank money certificates are not associated with commodity money even at the international level? Then the several money certificates float against one another, their relative values being determined by speculation on the foreign exchange market. As explained in section 10.1, the value of money certificates each day is a proportion of their value the day before, regressing back to when they were still associated with commodity money, the value of that commodity money regressing back to when it was valued primarily for its use value. When banking was still done at the community level and money certificates were all associated with commodity money, the day-to-day speculation which determined the value of money was primarily done by tradesmen, most of whom had little understanding of economics. But that was sufficient, because the speculation was on slowly-moving parameters such as the assimilation of isolated communities. As the nationalization of banks disassociated money certificates from commodity money, the foreign exchange market grew apace, attracting professional speculators who knew what it meant when, say, a central bank increased its purchase of government bonds with checks written against itself. This speculation is an outgrowth of the arbitrage operations discussed in section 15.3. When one borrows low intending to loan high, one first uses the borrowed funds to buy the foreign currency which alone can be used to purchase credit instruments at the higher interest rate. Thus, depreciating currencies are sold to buy stronger currencies, hastening their relative price adjustment.

The existence of professional speculators in the foreign exchange market, especially ones using sophisticated data-processing equipment, implies that the degree in which people who use a certain money certificate are hurt by its devaluation is not necessarily proportional to their distance from the central bank that issues it. In a closed economy, one benefits from a loose monetary policy in proportion to one's proximity to the central bank. Not counting the overall wastage of capital, it is a zero-sum game. Nobody could be farther from a central bank than foreign suppliers. However, when professional speculators cause the exchange rate to adjust immediately to a domestic monetary expansion, it is as though all foreign suppliers had added a price-change premium to their prices when quoted in the depreciating currency. Thus, unlike domestic suppliers far from the central bank who do their own speculation on the value of their currency, foreign suppliers have their price-change premium added automatically. Countries that are very dependent on foreign phenomena find that lowering their interest rate by issuing money does not stimulate investment because the prices of foreign capital adjust so quickly.

Thus, monetary expansion in a floating exchange rate system discourages imports because of the price-change premium automatically added by foreign suppliers and encourages exports until the new money has diffused through the domestic economy, raising prices as it goes. This latter assertion assumes that domestic suppliers far from a central bank do not add a price-change premium as quickly as the exchange rate adjusts through professional speculation. By the Law of Price Adjustment, their prices adjust immediately as money diffuses through them, it just takes time to diffuse, which is why they are slower than professional speculators who do not wait for new money to appear but adjust the rate of exchange immediately upon hearing of a central bank purchasing government bonds. Mercantilists would have rejoiced at an excess of exports over imports: It filled their bank's coffers with commodity money and, while they may have clothed their arguments in talk about national wealth, they knew that it was commodity money and not the capital stock that paid soldier's wages. However, in a floating exchange rate system, there are no international movements of commodity money and, also, mechanized warfare has made a nation's capital stock more decisive than its ability to hire soldiers. Today, most people support a favorable balance of trade because they resent foreign-owned capital projects in their country. Whatever the modern motivation for promoting exports over imports may be, it should be pointed out that monetary expansion promotes exports only by short-changing domestic suppliers, either by inducing them to unwittingly make a gift of part of their capital stock to foreign consumers or by inducing them to work for a lower real wage than would otherwise satisfy them. This may not be what advocates of currency devaluation have in mind when they speak of increasing a nation's wealth by encouraging exports.

The only thing accomplished by lowering the interest rate through monetary expansion is to discourage savings, especially from foreigners who only buy short-term credit instruments. Thus, the foreign savings on which small countries are dependent is more sensitive to their monetary policy than is the domestic savings on which large, isolated countries depend. Since small countries are usually not very enthusiastic about direct investment by foreigners, they should be careful not to discourage foreign purchase of credit instruments. If they are too poor to finance their own capital projects, their only choice is between working for foreigners and selling them credit instruments.

That monetary expansion under a fixed and a floating exchange rate system discourages and encourages net exports, respectively, is not always made clear in the literature. Now the difference between fiscal and monetary policy should be clear: Fiscal policy (paying enough interest on government bonds to make them competitive with private bonds) discourages net exports indirectly by driving up the interest rate, or directly by the government borrowing from foreigners to buy their phenomena. Monetary policy (selling low-yield government bonds to the central bank) encourages net exports when domestic suppliers far from the central bank are slower to raise their prices than the foreign exchange rate adjusts, and by discouraging foreign savings. But it is my assertion that government consumption is wasted either way, so the trade balance is not of great importance, though variance among the several countries in their emphasis on fiscal or monetary policy makes for more phenomena being transported around the world than there would be otherwise. Generally, governments of large countries have a choice between fiscal and monetary policy while those of small countries have trouble financing their operation with either method. The former method fails because no one loans them money at a reasonable interest rate, and the latter method fails because money diffuses through their economies so quickly that few are taken unaware for long, especially if people are used to speculating on the value of their currency when setting prices.

Just as when money certificates are associated with commodity money, the central bank of a small country with a brisk import-export business is more limited than one of a large, isolated country. It is basic that no institution can eliminate the problem of scarcity. Eliminating one institution, such as honoring one's money certificates, does not deliver one from scarcity, for another institution, such as professional speculation in foreign currencies, arises. Phenomena has value only if it is scarce, and if one wants to keep some, one has to fight for it. In the case of central banks, that means taking and keeping a large enough country that it can be self-sufficient and can support the troops necessary to close the borders. This is difficult because the deployment of troops does not benefit from economies of scale. It is not hard to control a neighborhood, but it is a real feat to exercise the same degree of control over a city, and, despite popular opinion about organized crime, the connection between cities is tenuous. At the more legitimate level of national governments, they rarely have the strength to monopolize the issue of money certificates unless people can be convinced that it is necessary or inevitable.

The governments of large countries alternate between fiscal and monetary policy with enough regularity to have motivated a great deal of spurious economic research involving Fourier Analysis. There are several reasons for governments to change their policies, none of which indicate an inherent cycle in business activity. Agriculture, of course, cycles every year, but that does not cause a business cycle because it is anticipated.

1) Cycling between monetary and fiscal policy allows people to profit by cycling their assets between stocks and bonds, respectively. The potential profit makes "central bank watching" virtually all there is to being a financial expert. In general, however, only those who are actually told of impending changes in central bank policies can profit regularly. Because these people can cause, or at least encourage, changes in policy, central banks tend to act sporadically and to distance themselves from the public.

2) The central bank is the most powerful branch of government but, like the rest of government, it is a bureaucracy. It must continually tinker with the money supply to maintain the expert status that separates it from uneducated politicians. If the stock of money remained constant, people would inquire what the high salaries, staffs of research economists, and impenetrable jargon was for.

3) Public opinion is slow to organize itself and lags behind a change in policy. Thus, it often encourages the next change to be overly severe in the other direction.

4) If the government desires to nationalize industry, the best way to accomplish this is to have the central bank periodically flood the stock market with money to create a bubble. Then, when the bubble bursts they can criticize a "flawed" capitalist system prone to irrational exuberance. Since investors know that the central bank will continue to roll over the government's debt, government bonds are considered a safe place to put one's money during such times. This creates a large transfer of wealth from private industry to the government. With every cycle, more businesses are nationalized as they voluntarily come to the government asking to be bailed out.

Cycling helps wealthy people at the expense of poor people. This is because wealthy people's investments prosper greatly during booms and slightly during busts but they almost never actually regress. Poor people, however, do not have investments and are mostly indifferent to booms. During busts, a poor person may fall behind on his mortgage or car payments and lose an asset worth many times the value of the missed payments. Also, a continuous employment history is highly valued (because it demonstrates stability) and a gap of a few months or a year can do irreparable damage to one's career. This is especially true if one graduates from school during a bust because, if one's education does not get one a job within a year or two, it never will. Thus, many people who were middle class during one boom spend the next one at the bottom rung of a different career ladder and without a car. Similarly, if they fell behind on their car insurance payments during the bust, they may spend the next boom devoting a quarter of their income to the local police department's usurious fine schedule. Wealthy organizations like banks and police departments can easily tolerate occasionally stagnant returns on their investments because they make most of their money from confiscation and usury, which is dependent on the existence of a business cycle. Thus, those in power never give more than lip-service to the idea of steady growth.

The first task of a central bank is to cloud the subject of economics. As Appendix C demonstrates, mainstream economics could be condensed into a one-year course in applied mathematics. One need only require that students already be familiar with multivariable calculus. Instead, economics is granted departmental status, billions of dollars in research grants are spent to support it, and multivariable calculus is never required of economists at any level. The result is a lot of career economists who have never actually considered the three identities of mainstream economics as a nonlinear system of simultaneous equation. Once mainstream economics is described mathematically, then it is no more difficult than any number of similar problems in the physical sciences. For instance, the conservation of momentum and the conservation of energy are used by physicists to set up systems of two equations in two unknowns. Mainstream economics seems difficult only because its practitioners are not familiar with axiomatic systems. One need only recall Hume's discussion of billiard balls (see Section 11.2) to see the difficulties encountered when one does not begin a discussion by carefully stating one's axioms.

Research economists, whose only job is to write journal articles, make as much money by grant hunting as any professional in private industry does. Journal articles are about ten pages long and take for granted all of mainstream economics except the one obscure question that the article addresses. They are not the place to address fundamental questions of theory. Furthermore, it is impossible since one's first few publications must be co-signed by one's academic advisor. By the time one is tenured and acting as an advisor to the next generation of students, one has far too great a financial stake in the system to question its axioms or even to state them clearly. This makes the business of academics, otherwise known as "grant hunting", an extremely conservative one which is incapable of addressing the fundamental issues in a field. At best it can only refine existing theories, and then it often succeeds only in making them too obscure to be useful in practical situations. This is particularly true of economics because it is politically motivated, but it is also true in the hard sciences. As one of my physics professors once stated (after noting that he could only say such things because he was retiring), "If Michelson and Morley were alive today, they would have found the ether and applied for another $500,000 research grant to measure it."

In **The Crisis of Vision in Modern Economic Thought**, Heilbroner and Milberg write[21]:

> *Those who hold positions at the few "top" universities carry a disproportionate amount of power over hiring, publishing, and the granting of research funds, including federal government research funding through the National Science Foundation. The hiring of new economists by the leading universities is limited to those trained at only a few select graduate programs. And the major academic journals are equally dominated by articles written by economists from those same schools. Graduates of only seven departments accounted for 54%, 58%, and 74% of articles published in the* American Economic Review, *the* Journal of Political Economy, *and the* Quarterly Journal of Economics, *respectively, for the years 1973-8. Within such a structure, the possibilities for a fundamental change in the direction of thinking are obviously limited.*

No true advances in economics or any other scientific field can be made in the context of grant hunting.[†] Unfortunately, the system of federally-funded research is too well entrenched to be moved by public criticism and so one must focus on a more obtainable goal. Relieving economics of its departmental status and making it a course in applied mathematics would be a worthy and realistic project for student activists. Since many universities are experiencing budget constraints, such a reorganization could be presented to their administrations as a cost-cutting measure. Mathematics departments, which are currently given only a small fraction of the grant money given to economics departments, would welcome such a move. Only economists would be left out in the cold, as they could never compete with real mathematicians in the now wealthy mathematics departments. However, economists are given little respect by the public and none by mathematicians, so such a reorganization could probably be enacted over the protests of an economics faculty.

[†] I was never a grant hunter or aspired to be one. In rudimentary form, all 16 sections of this book existed in 1988 before I enrolled in college. Now I have a B.S. in mathematics and, like most of my friends in the math and engineering departments, I took economics to satisfy my school's core curriculum. I represent the engineers who sit in the back of economics classes playing computer chess, not the business majors who sit in the front row posing questions with ludicrous solemnity.

The five major mistakes of mainstream economics are:

1) Addressing supply and demand instead of price and stock;
2) Assuming that all short term credit instruments function as money;
3) Believing that the average price level is a meaningful statistic and, hence, that prices are "sticky;"
4) Believing in a "natural" rate of interest; and
5) Assuming "perfect" competition, that is, lots of tiny firms selling a homogenous product.

The first three are the principle mistakes of mainstream economics. Student activists should press their professors on these points as often as possible. The best way to defeat an entrenched dogma is to query professors on the more preposterous parts until they are forced to speak clearly and admit that it is all nonsense. The "paradox of thrift" certainly qualifies, and so would their business cycle theory if they had one. Business cycles take place over time, but mainstream economists insist on taking a snapshot of the economy and then analyzing it without any reference to the process that it is a part of. Thus it does not seem ironic to them that their policy recommendation during recessions is more government spending when it was government waste that caused the recession in the first place. One of the most preposterous statements that one will encounter in first-semester macroeconomics is the assertion that infinite increases in government spending cause infinite increases in Gross National Product. This "multiplier effect" is an easy corollary of the one-dimensional model called the "Keynesian Cross." Appendix C gives a detailed explanation of mainstream macroeconomics, which is actually three-dimensional, through economists have only proven capable of considering two dimensions at a time. Most economists are not aware that it is possible to solve three equations in three unknowns simultaneously. The lack of understanding of even lower-division mathematics (multi-variable calculus) in economics departments is appalling. This is sometimes concealed by the fashion in economics journals to fill the articles with complicated-looking but meaningless equations.

Microeconomics is even worse. The method of mainstream economics really has a third variable which is never mentioned and that is the time unit for supply and demand. It is well known that elasticity is a function of this time unit and, if this is true, one calculates a different price depending on whether one speaks of weekly or monthly supply and demand. This is an inconsistency since there can only be one price and it is not dependent on the caprice of an economist when he decides how often to conduct his surveys. This is a point which is glossed over in mainstream texts. A detailed discussion of the time unit chosen for supply and demand is never given and many texts neglect to mention the need for choosing one at all. Yet in their chapter on elasticity, every textbook lists time as a factor, sometimes as the most important factor.

Whenever supply and demand are invoked, students should ask their professor what time unit he is using, why he chose that time unit and why he thinks his price is the same and his quantities are proportional to other possible time units that he might have chosen. Also, ask him why certain individuals are on the demand curve (and are called "buyers") and others are on the supply curve (and are called "sellers") when either would buy or sell if the price were right. Mainstream economists have two variables, price and quantity per unit of (some usually unspecified) time, and two equations, supply and demand. For this to work at all, the equations must be independent, which means that each individual must be either a buyer or a seller. The economist's decision to put people on one curve or the other cannot depend on the price that they would buy or sell at because both equations are defined for all prices. (Price is one of the independent variables.) So what is the economist's decision based on? Ask him repeatedly until he admits that there is really only one distribution. Also, press him to acknowledge that the demand distribution independently exists at each instant of time. Supply and demand curves are different depending on the time unit chosen. Mainstream economists provide no proof that their predicted prices are independent of their choice of time unit. For example, will thirteen predicted weekly quantities be the same as three predicted monthly quantities?

A large part of the problem with supply and demand is that it is used descriptively, but called predictive. It is easy to predict the past. Economists just observe the quantity produced one month and what it sold for and they put a little × over that spot. Then, by pure conjecture, they draw four tails on their × to fill their graph paper. Supply and demand has never been used predictively, not even to make bad predictions. × marks the spot is a purely descriptive technique. Since they are using the 20-20 vision of hindsight, they can do this for three months in a row and, to nobody's surprise, the sum of the quantities is the quarterly quantity. In the real world, price is constant for years at a time but, for most companies, their weekly and monthly sales figures swing wildly and unpredictably, sometimes by several fold from one month to the next. Mainstream economists have no explanation for this, which they should since their theory is called supply and demand and the horizontal axis of their graph is labeled weekly (or monthly) quantity. When I have been asked to help predict sales, I have told them that price is related to stock, not supply, and that they should stop watching their sales chart so ardently. At most companies, there is someone in accounting who feeds sales figures to the employees so that they can predict layoffs. They know that every dip in sales will send hundreds of them to the unemployment office, and that every rise will have their bosses clapping each other on the back and extolling their brilliant and farsighted management. They also know that nobody can predict sales. Supply never means anything in economics, though sometimes (for non-durable phenomena) it can pass for stock.

Engineers who have taken a course in economics to satisfy their school's core curriculum have often wondered how economics could survive for decades while clearly being shot full of holes. This is largely because the government pays for all economic research and they get the theory they call for. However, while research economists live very comfortably at the taxpayer's expense, their colleagues who work for banks actually make more. Economic theory has no application to employment at a bank and banks are managed by people who are smart enough to see through the holes in mainstream economics. Yet they continue to insist on a master's degree or a Ph.D. in economics. Cynically, it has been suggested that this is because many banks are corrupt and they fear having a self-righteous employee blow the whistle on them. The best way to weed out such people is to ask them to sit through six or eight years of spurious theorizing. The virtuous will get disgusted and walk out while the more mercenary will keep their eyes on the carrot and bite their tongues whenever the urge to laugh at their professor arises. It is the latter group that the banks want to hire, for they will be privy to the fraud and extortion practiced by their employer and must not speak of it. Also, they need to age six or eight years before working at a bank and studying economics keeps them out of prison in the meantime. Furthermore, graduate studies leave them up to their necks in debt, which makes them loyal to their employer until they get that paid off.

It is time that we exploded the myths of mainstream economics. It is time that we installed an axiomatic theory, in the true mathematical sense of the word. Today, economics is like physics was in the 17th century. But nothing prevents finding a set of axioms for economics just as Newton found a set of axioms for physics.

Appendix A: Alternative Distributions for First-Unit Demand

In section 11.3 the question was raised: What if people had traditionally assumed that the time value of money is an harmonic function rather than an exponential function? The present value of money to be received a year from now is currently calculated by multiplying the amount by a constant near unity called the common ratio. Under the proposed system, one would invert, add a constant, and invert again. By the Law of Proportionate Effect (Theorem 1, section 11.3), when the common ratio is taken to be not a constant but a random variable, first-unit demand is logrithmico-normally distributed (abbreviated lognormal). It is natural to make a similar inquiry into the distribution of first-unit demand when the value of money progresses harmonically:

$$m_j = \frac{1}{\dfrac{1}{m_{j-1}} + \varepsilon_j}$$

Each step is the last one inverted, added to a random quantity, and inverted again.

$$\frac{m_{j-1} - m_j}{m_j m_{j-1}} = \varepsilon_j$$

Solve for ε_j.

$$\sum_{j=1}^{n} \frac{m_{j-1} - m_j}{m_j m_{j-1}} = \sum_{j=1}^{n} \varepsilon_j$$

Find the sum of all ε_j from the initiation of the process to its termination after n steps.

$$\int_{m_0}^{m_n} \frac{-dm}{m^2} = \sum_{j=1}^{n} \varepsilon_j$$

If each step is small, $m_j - m_{j-1}$ can be approximated by dm.

$$\frac{1}{m_n} - \frac{1}{m_0} = \sum_{j=1}^{n} \varepsilon_j$$ Integrate from m_0 to m_n.

$$\frac{1}{m_n} = \frac{1}{m_0} + \varepsilon_1 + \ldots + \varepsilon_n$$ Solve for $\frac{1}{m_n}$.

Here, and throughout this appendix, the random variables associated with each day's change in value, $\{\varepsilon_1, \varepsilon_2, \ldots, \varepsilon_n\}$, are taken from an unspecified distribution, but one with a finite mean and a non-zero, finite variance. This assumption is included in the characteristics of proportionate effect, in the characteristics of harmonic effect (each day's value changes by inverting, adding a random variable, and inverting again), and in any other axiom considered for the genesis of first-unit demand.

As can be seen from the last step, the reciprocal of one's indifference point after the n'th day is a constant (the reciprocal of its initial quantity) with a large number of random and identically-distributed quantities accumulated onto it. Hence, after having lived through n days and having seen their point of indifference change by a small amount each day, consumers of their first unit are normally distributed with regard to the variable $\frac{1}{m}$ and, hence, are reciprico-normally distributed with regard to the variable m. This relation shall be known as the Law of Harmonic Effect.

I will call reciproco-normally distributed phenomena ones whose reciprocals are normally distributed. The distribution is:

$$c_0(m) = \frac{e^{-\frac{1}{2}\left(\frac{\frac{1}{m} - \mu}{\sigma}\right)^2}}{\sqrt{2\pi\sigma m^2}}.$$ To save a division (divisions are costly in computer time), the linear transformation can be written $\frac{1 - \mu m}{\sigma m}$. The reciproco-normal distribution is illustrated in figure 15.

Figure 15

The Reciprico-Normal Distribution

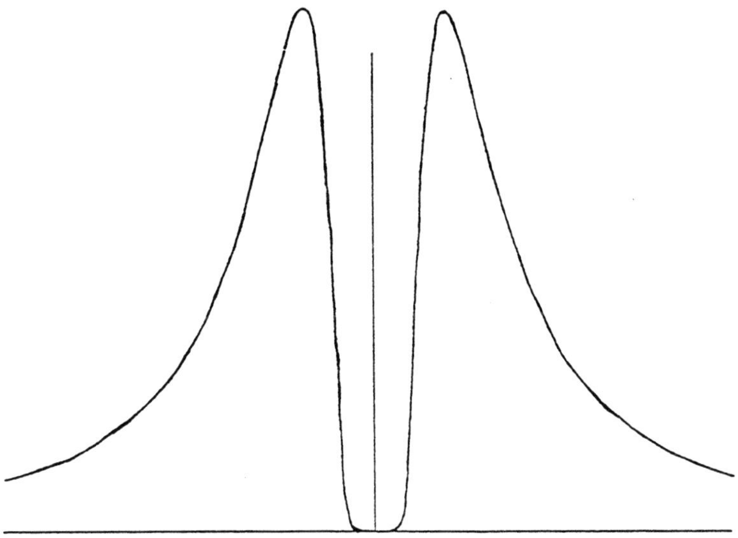

The properties of the arithmetic, geometric, and harmonic progressions are well known, though they are usually studied in isolation of one another. I propose that these are the first three of a sequence of progressions and, correspondingly, the normal, lognormal, and reciproco-normal distributions are the first three of a sequence of distributions.

Consider phenomena on which each day's events add a small quantity to that phenomena's value. If that quantity is always the same, it is called the common difference and the day-to-day values for that phenomenon follow an arithmetic progression. If that quantity is a random variable, in time the values of many such phenomena will be normally distributed. Notice that the quantity that is normally distributed, x, is an anti-derivative of 1, which may also be written x^0.

Consider phenomena on which each day's events multiply that phenomena's value by a quantity near unity. If that quantity is always the same, it is called the common ratio and the day-to-day values for that phenomenon follow a geometric progression. If that quantity is a random variable, in time the values of many such phenomena will be lognormally distributed. Notice that the quantity that is normally distributed, $\ln(x)$, is an anti-derivative of x^{-1}.

Consider phenomena on which each day's events invert that phenomena's value, add a small quantity, and invert again. If that quantity is always the same, the day-to-day values for that phenomenon follow an harmonic progression. If that quantity is a random variable, in time the values of many such phenomena will be reciproco-normally distributed. Notice that the quantity that is normally distributed, $\frac{1}{x}$, is an anti-derivative of $-x^{-2}$.

It is easy to see the pattern: Let an anti-derivative of x^0, x^{-1}, x^{-2},... be normally distributed and one has defined a sequence of distributions. The derivation of the associated sequence of progressions is as follows:

$$m_j = \frac{1}{\frac{1}{m_{j-1}} + m_{j-1}{}^k \varepsilon_j}$$

Each step is the last one inverted, added to a random quantity (which is multiplied by the k'th power of the last step), and inverted again.

$$\frac{m_{j-1} - m_j}{m_j m_{j-1}{}^{k+1}} = \varepsilon_j$$

Solve for ε_j.

$$\sum_{j=1}^{n} \frac{m_{j-1} - m_j}{m_j m_{j-1}{}^{k+1}} = \sum_{j=1}^{n} \varepsilon_j$$

Find the sum of all ε_j from the initiation of the process to its termination after n steps.

$$\int_{m_0}^{m_n} \frac{-dm}{m^{k+2}} = \sum_{j=1}^{n} \varepsilon_j$$

If each step is small, $m_j - m_{j-1}$ can be approximated by dm.

$$\frac{1}{(k+1)m_n^{k+1}} - \frac{1}{(k+1)m_0^{k+1}} = \sum_{j=1}^{n} \varepsilon_j$$

Integrate from m_0 to m_n.

$$\frac{1}{m_n^{k+1}} = \frac{1}{m_0^{k+1}} + (k+1)(\varepsilon_1 + \ldots + \varepsilon_n)$$

Solve for $\frac{1}{m_n}$.

As can be seen from the last step, the reciprocal of one's indifference point after the n'th day raised to the (k+1)'th power is a constant (the reciprocal of its initial quantity raised to the (k+1)'th power) with a large number of random and identically-distributed quantities accumulated onto it. Hence, after having lived through n days and having seen their point of indifference change by a small amount each day, consumers of their first unit are normally distributed with regard to the variable $\frac{1}{m^{k+1}}$. This relation shall be known as the Generalized Law of Harmonic Effect.

The generalized harmonic progression is given by $m_j = \dfrac{1}{\dfrac{1}{m_{j-1}} + m_{j-1}{}^{k}\varepsilon_j}$. This is the harmonic progression when k equals 0 and, for k = 1, 2, 3,..., one gets the latter progressions in the sequence. The distribution associated with the k'th progression is:

$$c_0(m) = \frac{(k+1)e^{-\frac{1}{2}\left(\frac{\frac{1}{m^{k+1}}-\mu}{\sigma}\right)^2}}{\sqrt{2\pi}\sigma m^{k+2}}$$

With the invention of any new distribution, a number of theorems must be proven regarding its mean, its variance, and a host of other characteristics used to describe statistical distributions. Also, an investigation must be made regarding what types of phenomena are modeled by the distribution. In the case of the reciprico-normal distribution, that means asking what types of phenomena change each day by inverting their current value, adding a random quantity, and inverting again. As I have invented not only the reciprico-normal distribution but an entire sequence of distributions, each with an associated progression, this is a fairly daunting task. However, the purpose of this appendix is not to introduce new research in pure mathematics, so the investigation of this sequence of distributions will be deferred for separate treatment. The purpose of this appendix is to attack my third axiom, that is, to attack my choice of the lognormal distribution for first-unit demand. Here, we will be content to address one application: Can the reciprico-normal distribution or a later distribution in the sequence be used to describe first-unit demand? If not, what steps must be taken to find an alternative to the lognormal distribution and are they executable?

The harmonic progression cannot be used to calculate the time value of money because its summation does not converge; that is, to replace our third axiom with the assertion that first-unit demand conforms to the characteristics of harmonic effect would imply that the present value of money held at interest is always infinite. Therefore, the reciproco-normal distribution is not a candidate for first-unit demand and, for the same reason, neither is the normal distribution. To test the latter distributions in the sequence, one must determine whether their associated progressions converge or diverge.

First, let us explicitly prove that the harmonic progression diverges: $f(t) = \dfrac{1}{c + \varepsilon t}$ diverges because $\displaystyle\int_1^{\infty} \frac{dt}{c + \varepsilon t} = \left.\frac{\ln(c + \varepsilon t)}{\varepsilon}\right|_1^{\infty} = \infty$ with c and ε positive constants. By the Integral Test for convergence, $\displaystyle\sum_{t=1}^{\infty} f(t) = \infty$. Now consider the value of this function one day after an arbitrary time, t. $f(t+1) = \frac{1}{c+\varepsilon(t+1)} = \frac{1}{c+\varepsilon t+\varepsilon} = \frac{1}{\frac{1}{f(t)}+\varepsilon}$ This last step is the harmonic progression from which the reciproco-normal distribution is derived. Thus, if $t = 0, 1, 2,\ldots$, then $f(t)$ parametrically defines the same ordered set of numbers that is defined recursively by the harmonic progression. Since $f(t)$ diverges, the harmonic progression diverges.

Notice that the smaller ε is, the faster $f(t)$ diverges. This is because, for every t, as ε tends toward zero the numerator of $\frac{\ln(c+\varepsilon t)}{\varepsilon}$ approaches $\ln(c)$, which is constant, and the denominator approaches zero.

Now consider the k'th progression in the sequence. Defined recursively, $f_k(t+1) = \frac{1}{\frac{1}{f_k(t)}+\varepsilon f_k(t)^k}$ with $\varepsilon > 0$. $f_k(t)$ is negative monotonic and it approaches zero as t approaches infinity, so there exists T such that, for all $t > T$, $f_k(t)^k < f_k(t) < 1$ and, hence, the k'th progression diverges faster than the harmonic progression. Thus, of the sequence of distributions beginning with the normal, lognormal, and reciprico-normal, only the lognormal qualifies for the distribution of first-unit demand because only the geometric progression's summation converges.

Let us explicitly list the steps that must be taken to provide an alternative to the axiom that first-unit demand conforms to the characteristics of proportionate effect, that is, an alternative to the lognormal distribution:

1) Choose a negative-monotonic function $f:\Re^+ \rightarrow \Re^+$ such that $\lim_{t\rightarrow\infty} f(t) = 0$.

2) Prove that the integral of $f(t)$ converges, that is, $\int_1^\infty f(t)dt < 0$.

3) Find the inverse (in composition), $f^{-1}(m)$, and set $m_j = f(f^{-1}(m_{j-1}) + \varepsilon_j)$. $f^{-1}(m)$ must be defined over all $\Re^+$.

4) Separate (3) so $g(m_j, m_{j-1})(m_j - m_{j-1}) = h(\varepsilon_j)$ with $h(\varepsilon_j)$ monotonic.

5) Anti-differentiate $g(m)$ to get $G(m)$ after making the approximations $m = m_j$ and $m = m_{j-1}$.

6) By applying the Central Limit Theorem, define the alternative distribution to be $c_0(m) = g(m)\dfrac{e^{-\frac{1}{2}\left(\frac{G(m)-\mu}{\sigma}\right)^2}}{\sqrt{2\pi}\sigma}$, scaled so that the area under it is unity.

Any of these steps may fail, in which case a new candidate must be chosen. Specifically, the integral of $f(t)$ may not converge, $f(t)$ may not be invertible, (3) may not be separable as described, or $g(m)$ may not have an anti-derivative. If a function meets all of these criteria except possibly (2), I will call the associated distribution "paranormal." Paranormal distributions for which the summation of the associated progression converges are demand distributions. It is an easy exercise to prove that stock is always finite regardless of which demand distribution is chosen.

First, let us consider the exponential function, $f(t) = e^{-t}$, to show how it produces the lognormal distribution:

1) $f(t) = e^{-t}$

2) $\displaystyle\int_1^\infty e^{-t}dt = -e^{-t}\Big|_1^\infty = e^{-1} < \infty$

3) $f^{-1}(m) = -\ln(m)$ with $m > 0$

Thus, $m_j = e^{-(-\ln(m_{j-1}) + \varepsilon_j)} = \dfrac{m_{j-1}}{e^{\varepsilon_j}}$

4) $m_j = \dfrac{m_{j-1}}{e^{\varepsilon_j}}$

$$m_j - m_{j-1} = m_{j-1}\left(e^{-\varepsilon_j} - 1\right)$$

$$\frac{m_j - m_{j-1}}{m_{j-1}} = e^{-\varepsilon_j} - 1$$

5) $g(m) = \dfrac{1}{m}$ so $G(m) = \ln(m)$

6) $$c_0(m) = \frac{e^{-\frac{1}{2}\left(\frac{\ln(m) - \mu}{\sigma}\right)^2}}{\sqrt{2\pi}\sigma m}$$

Now let us consider the p-series with p = 2. Two is a natural choice for p because $\sqrt{m_{j-1}} - \sqrt{m_j}$ has a conjugate.

1) $f(t) = \dfrac{1}{t^2}$

2) $$\int_1^\infty \frac{dt}{t^2} = -\frac{1}{t}\bigg|_1^\infty = 1 < \infty$$

3) $f^{-1}(m) = \dfrac{1}{\sqrt{m}}$ with $m > 0$

Thus, $$m_j = \frac{1}{\left(\frac{1}{\sqrt{m_{j-1}}} + \varepsilon_j\right)^2}$$

4) $$m_j = \frac{1}{\left(\frac{1}{\sqrt{m_{j-1}}} + \varepsilon_j\right)^2}$$

$$\frac{\sqrt{m_{j-1}} - \sqrt{m_j}}{\sqrt{m_j m_{j-1}}} = \varepsilon_j$$

$$\frac{\sqrt{m_{j-1}} - \sqrt{m_j}}{m_{j-1}\sqrt{m_j} + m_j\sqrt{m_{j-1}}} = \varepsilon_j$$

5) $$g(m) = \frac{-m^{-\frac{3}{2}}}{2} \qquad \text{so } G(m) = \frac{1}{\sqrt{m}}$$

6) $$c_0(m) = \frac{e^{-\frac{1}{2}\left(\frac{\frac{1}{\sqrt{m}} - \mu}{\sigma}\right)^2}}{\sqrt{2\pi}\sigma m^{\frac{3}{2}}}$$

This proves that there is an alternative to the lognormal distribution for first-unit demand. I will call it the root-reciprico-normal distribution. It is similar in appearance to the lognormal distribution but it converges much more slowly. By converges more slowly I mean that, regardless of their respective linear transformations, the root-reciprico-normal distribution eventually dominates the lognormal distribution. By way of comparison, without any linear transformation ($\mu = 0$ and $\sigma = 1$) 99% of the probability in the lognormal distribution is from 0 to 10, while one must integrate the root-reciprico-normal distribution from 0 to 6400 to get 99% of its probability.

Hence, an entirely independent theory of economics could be constructed on the basis of the following three axioms:

1) One's value scale is totally (linearly) ordered:

- **i)** Transitive; $p \leq q$ and $q \leq r$ imply $p \leq r$
- **ii)** Reflexive; $p \leq p$
- **iii)** Antisymetric; $p \leq q$ and $q \leq p$ imply $p = q$
- **iv)** Total; $p \leq q$ or $q \leq p$

2) Marginal (diminishing) utility, $u(s)$, is such that:

- **i)** It is independent of first-unit demand.
- **ii)** It is negative monotonic; that is, $u'(s) < 0$.
- **iii)** The integral of $u(s)$ from zero to infinity is finite.

3) First-unit demand conforms to proportionate effect:

- **i)** Value changes each day by taking the square root of the previous day's value, inverting, adding a quantity (called ε_j, with j denoting the day), squaring, and inverting again.

ii) In the long run, the ε_j's may be considered random as they are not directly related to each other nor are they uniquely a function of value.

iii) The ε_j's are taken from an unspecified distribution with a finite mean and a non-zero, finite variance.

My choice of proportionate effect for first-unit demand (and, hence, my use of the lognormal distribution) has a solid intuitive basis because it is the generalization of calculating the interest owed on a debt per unit of time as a percentage of the amount owed (see section 11.3). Since this is how people have calculated interest throughout recorded history, the weight of tradition supports my choice of the lognormal distribution. It is unlikely that critics will succeed in convincing the banking industry to calculate the interest owed on a debt per unit of time by taking the square root of the amount owed, inverting, subtracting a constant, squaring, and inverting again. However, it is interesting that alternative systems exist which contain no internal contradictions but which are still very different from human economies. Perhaps the inhabitants of another planet organize their economy in this or another way, and consider their calculation of interest to be as natural as we consider multiplication by a constant.

In the proposed system, the value of money held at interest would grow considerably faster than exponentially. From step four (above), $m_j - m_{j-1} = m_{j-1}\left(e^{-\varepsilon_j} - 1\right)$ implies that the rate of growth for money held at interest is $\frac{dm}{dt} = \mathrm{k}m$ when first unit demand is modeled by the lognormal distribution. $\sqrt{m_{j-1}} - \sqrt{m_j} = \left(m_{j-1}\sqrt{m_j} + \mathrm{m}_j\sqrt{m_{j-1}}\right)\varepsilon_j$ implies that the rate of growth for money held at interest is $\frac{dm}{dt} = -2\mathrm{k}m^{\frac{3}{2}}$ when first unit demand is modeled by the root-reciprico-normal distribution. Here k must be a negative constant to model savings rather than depreciation, which is why I said "subtract a constant" in the previous paragraph. This latter differential equation is separable and has a solution, $m = \mathrm{k}^{-2}t^{-2}$, but that is not the solution we want. I do not know an explicit formula for accumulated compound interest under the proposed system. Because it dominates exponential growth, one cannot use a polynomial approximation. One might consider using the sequence e_0, e_1, e_2,... that I developed in section 12.1 or some similar method. I leave the approximation of a positive monotonic function $m(t)$ such that $\frac{dm}{dt} = -2km^{\frac{3}{2}}$ for $k < 0$ as an exercise for the reader.

If anything, critics of our system (exponential growth of money held at interest) should look for a function that meets all of the criteria listed above but for which the value of money held at interest grows less than exponentially. I am not aware that anyone has ever criticized the technique of multiplying by a constant or proposed any other formula, but some have been shocked by how much interest can accumulate over long periods of time. If a boy deposits some money in a bank and then returns for it as an old man, the bank will probably conjure up some rule about not paying off dormant accounts. This is partly because they can but also because the bank feels that the time value of money is accurately modeled by the exponential function for only a short time. People who are preparing to purchase a house with a 30-year mortgage are told how much their monthly payments will be but never the total amount of money that they will spend over the next 30 years. Most people would be appalled at the huge sum of money that they are going to spend relative to the cash price of their house. Thus, the proposed system, which dominates exponential growth, is not a practical alternative to exponential growth for money held at interest and, concomitantly, the root-reciprico-normal distribution is not a practical alternative to the lognormal distribution. I leave it as an exercise for the reader to find a practical alternative to the lognormal distribution. I would be interested in hearing from any of my readers who can successfully attack my third axiom or, for that matter, any of my axioms.

Appendix B: BASIC Program Listing

I

Overview of the Software

Following is the source code for a program written in QBasic which will print out the demand distribution with the saturation point marked by a rectangle. Recall that the saturation point is defined so that the area of the rectangle formed by the two axis and horizontal and vertical lines extending from the point *m*,c(*m*) is equal to the tail area, which represents stock. There are three parameters which affect the position of the saturation point:

1) u(*s*), marginal (diminishing) utility;

2) μ, importance of the phenomenon relative to money; and

3) σ, difficulty of substituting other phenomena.

μ and σ are input at run time while u(*s*) is input by retyping its definition. QBasic is interpretive, so retyping a line is not as inconvenient as it is with a compiled language. Readers may wish to vary each parameter independently so that they can see its effect on saturation and the shape of the distribution. μ must be greater than or equal to zero, σ must be greater than zero, and u(*s*) must be a negative-monotonic probability density function. Explanations of the algorithms used in this program are given in section 12.1.

If one is restricted to the use of a programmable calculator, one can translate the "initial estimate" section of the BASIC program into the calculator's language and use the initial estimate of the saturation price without further refinement. It is accurate enough to give the reader an idea of how the saturation price reflects changes in the three parameters; u(*s*), μ, or σ. This works particularly well if one defines

$u(s) = e^{\ln(k)s}$, with k input at run time. One can set G at k and omit the FOR-NEXT loop used to calculate G.

II

An Upper Bound for Multiple Saturation Prices

I have not found any set of parameters μ, σ, u(s) such that the following assertion fails. However, the inequality check-marked (below) can be made to fail with some u(s) functions, so the assertion is only informally justified, not proven. Monotonicity is stronger than necessary, however, and when it fails, it seems to do so at higher prices than any saturation point.

Assertion: There is at most one saturation price greater than $e^{\mu-\sigma^2\sqrt{1+\sigma^{-2}}}$.

Informal Justification: By Theorem 11, $e^{\mu+\sigma^2}$ is an upper bound for any saturation price so, if the second derivative is positive monotonic for some $m \le e^{\mu+\sigma^2}$, it cannot alternate from negative to positive and back to negative again in that region. The third derivative of aggregate utility is

$$\frac{1}{m}\sum_{r=0}^{\infty} c_0(x)\left(1 - \left(\frac{\ln(x)-\mu}{\sigma^2}\right)^2 + \frac{1}{\sigma^2}\right) \qquad \text{with } x = \frac{m\mathrm{u}(0)}{\mathrm{u(r)}}$$

To prove monotonicity of the second derivative, we must show that the third derivative is strictly positive in an interval extending up to $e^{\mu+\sigma^2}$. $c_0(x)$ and $\frac{1}{m}$ are positive for all x, so we only have to examine $1 - \left(\frac{\ln(x)-\mu}{\sigma^2}\right)^2 + \frac{1}{\sigma^2}$. This is positive for all

$e^{\mu-\sigma^2\sqrt{1+\sigma^{-2}}} < x < e^{\mu+\sigma^2\sqrt{1+\sigma^{-2}}}$. Let N be the greatest integer such that $e^{\mu+\sigma^2}\frac{u(0)}{u(N)} < e^{\mu+\sigma^2\sqrt{1+\sigma^{-2}}}$. Then the following inequality would show that the second derivative is positive monotonic from $e^{\mu-\sigma^2\sqrt{1+\sigma^{-2}}}$ to $e^{\mu+\sigma^2}$.

$$\frac{1}{p}\sum_{r=N+1}^{\infty} f_0''\left(p\frac{u(0)}{u(r)}\right) > -\frac{1}{p}\sum_{r=0}^{N} f_0''\left(p\frac{u(0)}{u(r)}\right)$$

with $f_0''(x) = c_0(x)\left(1 - \left(\frac{\ln(x)-\mu}{\sigma^2}\right)^2 + \frac{1}{\sigma^2}\right)$ and $p = e^{\mu+\sigma^2}$

Unfortunately, without imposing arbitrary restrictions on u(s), it can be constructed so that the inequality check-marked (below) fails, and by enough that the latter two inequalities in the sequence do not make up for it. However, for less artificial u(s) functions, the following inequalities support the assertion.

$$\frac{1}{p}\sum_{r=N+1}^{\infty} f_0''\left(\frac{pu(0)}{u(r)}\right) > \int_{e^{\mu+\sigma^2\sqrt{1+\sigma^{-2}}}}^{\infty} f_0''(t)dt \qquad \checkmark$$

$$= c_0(t)\left[\frac{\ln(t)-\mu}{\sigma^2} - 1\right]_{\mu+\sigma^2\sqrt{1+\sigma^{-2}}}^{\infty}$$

$$= \frac{1-\sqrt{1+\sigma^{-2}}}{\sigma e^{\mu+\sigma^2\left(\frac{1}{2}+\sqrt{1+\sigma^{-2}}\right)+\frac{1}{2}}}$$

$$\geq -\frac{1}{\sigma^3 e^{\mu+\frac{3\sigma^2}{2}}}$$

$$= -f_0''(p)$$

$$\geq -\sum_{r=0}^{N} f_0''\left(p\frac{u(0)}{u(r)}\right) \quad \blacksquare$$

The BASIC program does a linear search for relative maximas from zero to $e^{\mu+\sigma^2}$. By Theorem 11, $e^{\mu+\sigma^2}$ is an upper bound for any saturation price. Readers may use this program to look for a counter-example to the above assertion. I would be interested in hearing from any of my readers who can provide a counter-example or, if the assertion really is true for all combinations of μ, σ, and $u(s)$, any who can provide a proof.

After the program listing are seven example runs. The first three fix σ and set μ at 0, 1 and 2. Notice that changing μ has no effect on the shape of the graph or the relative position of the saturation point. It also has no effect on the stock (the area of the rectangle, which equals the area from the rectangle out to infinity), which is 2.36621 in all three graphs. The only thing affected is the saturation price (the horizontal side of the rectangle), which is a simple function of μ, $.7183068e^{\mu}$. This illustrates the Law of Price Adjustment discussed in section 13 and is an unexpected result of my choice of the lognormal distribution. My choice of proportionate effect for first-unit demand (and, hence, my use of the lognormal distribution) was made because it is the generalization of calculating the interest owed on a debt per unit of time as a percentage of the amount owed. Only years later, in college, did I have the computer resources to discover the Law of Price Adjustment, and the mathematical ability to prove it. The latter four graphs demonstrate the effects of varying σ.

BASIC program listing

```
REM Copyright by Victor Aguilar, all rights reserved
REM program written in QBASIC under MS-DOS
DECLARE FUNCTION equal% (x, y)
DECLARE FUNCTION area! (z) : DECLARE SUB fail (msg$)
COMMON SHARED black%, blue%, green%, cyan%, red%, magenta%, brown%, white%
COMMON SHARED big%, top%, bottom%, left%, right%, tolerance
black% = 0: blue% = 1: green% = 2: cyan% = 3
red% = 4: magenta% = 5: brown% = 6: white% = 7
top% = 330: bottom% = -150: left% = -10: right% = 630
big% = 999: tries% = 99: DIM idu(big%), p(tries%), c%(right%)
SCREEN 12: VIEW:  ' Must use a VGA monitor -- 640x480, 16 color
WINDOW (left%, top%)-(right%, bottom%): COLOR cyan%

REM Input coefficients, find diminishing utility, and estimate the price
CLS : LOCATE 23, 4: PRINT "Quantify the following:"
INPUT "Importance of the phenomenon relative to money (1)          "; mu
INPUT "Difficulty of substituting other phenomena (.25)            "; sg
IF mu < 0 THEN fail "Importance must be non-negative."
IF sg <= 0 THEN fail "Substitutability must be positive."
isg = 1 / sg
INPUT "Common ratio for an exponentially diminishing utility (.5) "; k
COLOR white% + 8: tolerance = .000001 ' Relative accuracy of output
DEF fnu (s) = EXP(LOG(k) * s)
u0 = fnu(0): idu(0) = 1: g = 1
FOR n% = 1 TO big%
    un = fnu(n%): t = u: u = u + un: idu(n%) = u0 / un
    gn = idu(n% - 1) / idu(n%): g = g * gn
    IF gn >= 1 THEN fail "Common ratio must always be less than one."
    IF gn <= 0 THEN fail "Common ratio must always be positive."
    IF t = u THEN EXIT FOR
    NEXT n%: IF n% >= big% THEN fail "Utility did not converge."
e1 = EXP(sg): e2 = EXP(sg * sg)
m(0) = .4476505 - 8.923796E-02 * e1 + .4244496 * e2
m(1) = .4867678 - .0949248 * e1 + .3934986 * e2
m(2) = .5510112 - .1388672 * e1 + .3483012 * e2
m(3) = .3653524 - .0758399 * e1 + .2993362 * e2
g = g ^ (1 / n%): i% = INT(4 * g): p(0) = EXP(mu) * m(3)
IF g < .75 THEN
    p(0) = EXP(mu) * (m(i%) + 4 * (m(i% + 1) - m(i%)) * (g - .25 * i%))
END IF

REM Find the price, density, and stock at saturation.
FOR i% = 1 TO tries%
    f = 0: f1 = 0: f2 = 0
    FOR m% = 0 TO n%
        x = p(i% - 1) * idu(m%): z = (LOG(x) - mu) * isg
        z2 = z * z: c0 = EXP(-.5 * z2) / (x * sg): t = c: c = c + c0
        f1 = f1 + c0 * (z * isg - 1)
        f2 = f2 + c0 * (1 - z2 * isg * isg + isg * isg)
        IF t = c AND x >= EXP(mu + sg * sg) THEN EXIT FOR
        NEXT m%: IF m% > n% THEN m% = n%
    FOR r% = 0 TO m%
        z = (LOG(p(i% - 1) * idu(r%)) - mu) * isg
        f = f + area(z) / idu(r%): NEXT r%
    f = f - p(i% - 1) * c: h = f / (f1 - f * f2 / (2 * f1 * p(i% - 1)))
    p(i%) = p(i% - 1) - h: c = 0
    IF ABS(h / p(i%)) < tolerance THEN p = p(i%): EXIT FOR
    NEXT i%: IF i% >= tries% THEN fail "Newton's method did not converge."
FOR r% = 0 TO m%
    x = p * idu(r%): z = (LOG(x) - mu) * isg
    c = c + EXP(-.5 * z * z) / (x * sg): NEXT r%: pc = p * c
```

```
REM Fill c% array, the demand distribution.
PRINT "The price and stock at saturation are"; p; "and"; pc
scale = 256 ' Width of rectangle (height is one quarter of this.)
COLOR red% + 8
FOR p% = 1 TO right%: p0 = p% * p / scale: d = 0: f = 0
    LOCATE 28, 1: PRINT TAB(4); "percentage done:";
    PRINT USING "###%"; 100 * (p% / right%)
    FOR m% = 0 TO n%
        x = p0 * idu(m%): z = (LOG(x) - mu) * isg
        d0 = EXP(-.5 * z * z) / (x * sg): t = d: d = d + d0
        IF t = d AND x >= EXP(mu + sg * sg) THEN EXIT FOR
        NEXT m%: IF m% > n% THEN m% = n%
    FOR r% = 0 TO m%
        z = (LOG(p0 * idu(r%)) - mu) * isg
        f = f + area(z) / idu(r%): NEXT r%
        f = f - p0 * d
    IF f0 > 0 AND f <= 0 THEN
        max(k%) = p0 - (p * f / scale) / (f / f0): k% = k% + 1
        END IF
    IF d > c * 128000 / scale THEN
        c%(p%) = 32000
    ELSE
        c%(p%) = INT(.25 * scale * d / c + .5)
    END IF
    f0 = f: NEXT p%

REM List relative maximas and check assertion made in Appendix B
COLOR white% + 8: LOCATE 28, 1
PRINT "Aggregate utility is maximized near";
FOR k% = 0 TO 9
    PRINT max(k%); : b = EXP(mu - sg * sg * SQR(1 + isg * isg))
    IF max(k%) > b THEN max% = max% + 1
    IF max(k% + 1) > 0 THEN PRINT "and";  ELSE EXIT FOR
    NEXT k%: IF max(1) > 0 THEN PRINT  ELSE PRINT "only."
IF max% = 1 THEN
    PRINT "There is 1 relative maxima greater than"; b;
ELSE
    PRINT "There are"; max%; "relative maximas greater than"; b;
END IF

REM Print graph, axis, rectangle, etc.
FOR x% = 2 TO right%
    p% = x% - 1: LINE (p%, c%(p%))-(x%, c%(x%)), magenta% + 8
    NEXT x%
caxis% = green% + 8
LINE (0, top%)-(0, left%), caxis%  'Y axis doesn't go to bottom%
LINE (left%, 0)-(right%, 0), caxis%  'X axis
csat% = brown% + 8: xsat% = scale: ysat% = scale / 4  'saturation
LINE (0, ysat%)-(xsat%, ysat%), csat%
LINE (xsat%, 0)-(xsat%, ysat%), csat%
COLOR white% + 8: LOCATE 22, 29: PRINT p: LOCATE 20, 36: PRINT pc
COLOR red% + 8: LOCATE 2, 50: PRINT "Demand Distribution for"
COLOR brown% + 8: LOCATE 3, 50: PRINT "Axiomatic Theory of Economics"
COLOR red% + 8: LOCATE 4, 50: PRINT "by Victor Aguilar"
COLOR cyan%
END
```

```
FUNCTION area (z)
z2 = z * z: y = EXP(-.5 * z2): halfarea = 1.2533141373155#
IF ABS(z) < tolerance THEN
    a = halfarea
ELSE
    SELECT CASE z
        CASE IS > 1.28
            GOSUB contfrac
        CASE 0 TO 1.28
            GOSUB summation: a = halfarea - w
        CASE -1.28 TO 0
            z = -z: GOSUB summation: a = halfarea + w
        CASE IS < -1.28
            z = -z: GOSUB contfrac: a = 2 * halfarea - a
    END SELECT
END IF
area = a: EXIT FUNCTION

contfrac:
w = z * y: t = w
u = z: v = z2 + 1: s = y / u: m = s: a = w / v
FOR j% = 2 TO big%
    s = z * w + j% * y: y = w: w = s
    s = z * v + j% * u: u = v: v = s: s = m: m = a: a = w / v
    IF equal%(m, a) OR equal%(s, a) THEN EXIT FOR
    NEXT j%: IF j% >= big% THEN fail "Contfrac did not converge"
RETURN

summation:
w = z * y: t = w
FOR j% = 3 TO big% STEP 2
    t = t * z2 / j%: s = w: w = w + t
    IF equal%(w, s) THEN EXIT FOR
    NEXT j%: IF j% >= big% THEN fail "Summation did not converge"
RETURN

END FUNCTION

FUNCTION equal% (x, y)
IF x + y THEN
    equal% = ABS(2 * (x - y) / (x + y)) < tolerance
ELSE
    equal% = x = 0   'either x,y = 0 or x = -y
END IF
END FUNCTION

SUB fail (msg$)
CLS : LOCATE 15, 20: COLOR brown% + 8
PRINT "ERROR: "; msg$
END
END SUB
```

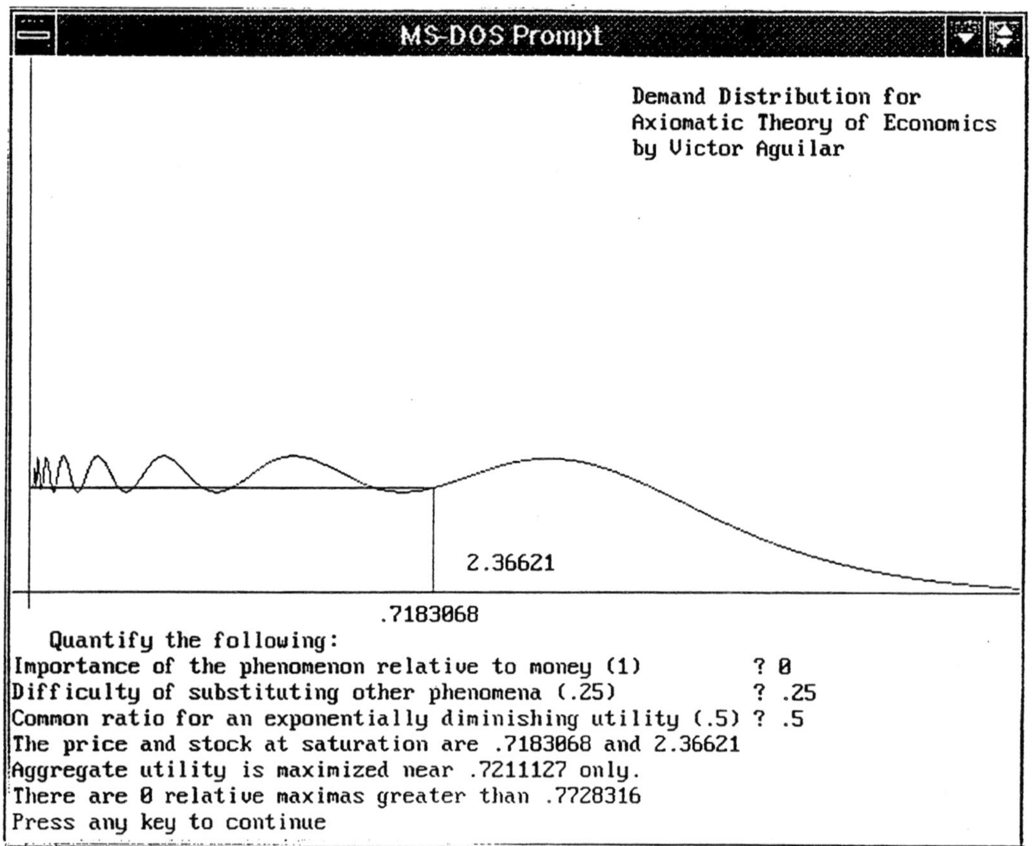
MS-DOS Prompt
Demand Distribution for
Axiomatic Theory of Economics
by Victor Aguilar
2.36621
.7183068
Quantify the following:
Importance of the phenomenon relative to money (1) ? 0
Difficulty of substituting other phenomena (.25) ? .25
Common ratio for an exponentially diminishing utility (.5) ? .5
The price and stock at saturation are .7183068 and 2.36621
Aggregate utility is maximized near .7211127 only.
There are 0 relative maximas greater than .7728316
Press any key to continue

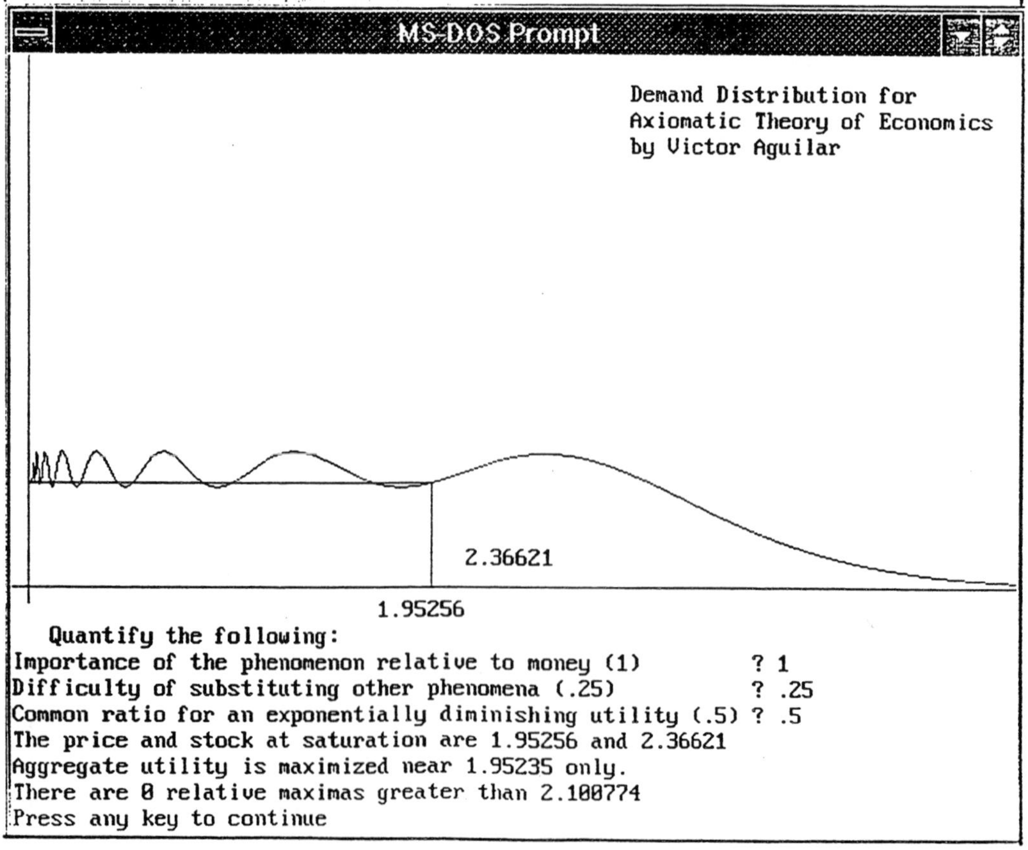
MS-DOS Prompt
Demand Distribution for
Axiomatic Theory of Economics
by Victor Aguilar
2.36621
1.95256
Quantify the following:
Importance of the phenomenon relative to money (1) ? 1
Difficulty of substituting other phenomena (.25) ? .25
Common ratio for an exponentially diminishing utility (.5) ? .5
The price and stock at saturation are 1.95256 and 2.36621
Aggregate utility is maximized near 1.95235 only.
There are 0 relative maximas greater than 2.100774
Press any key to continue

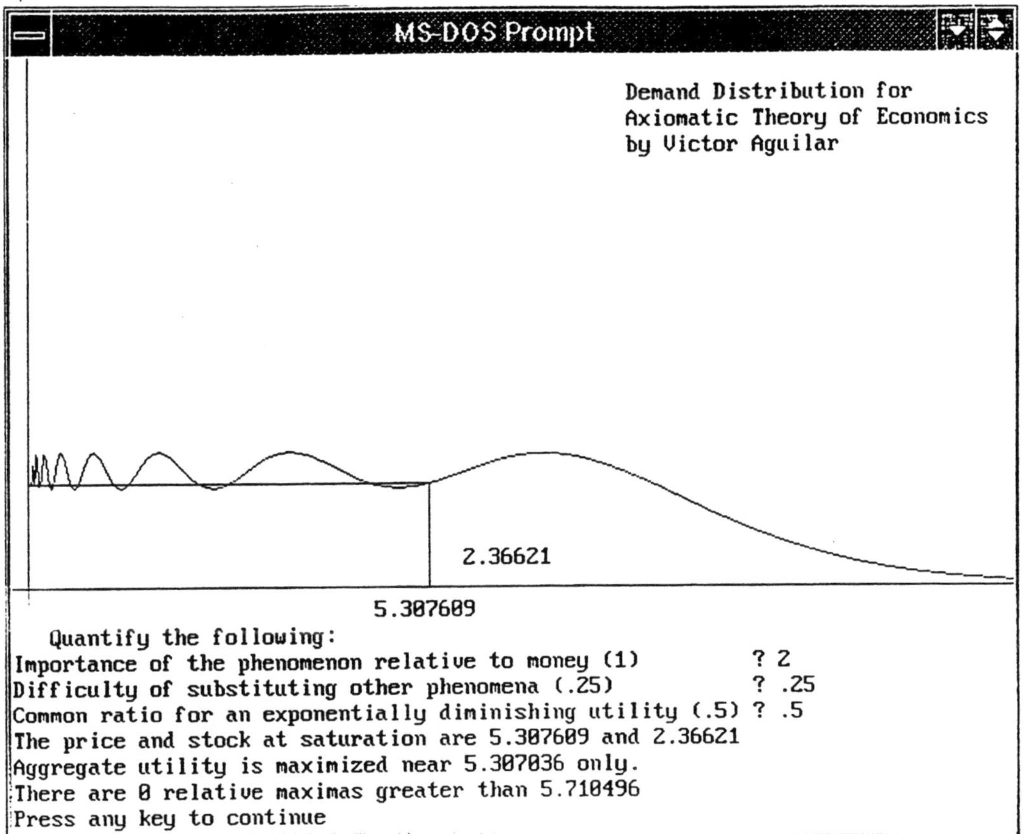
MS-DOS Prompt
Demand Distribution for
Axiomatic Theory of Economics
by Victor Aguilar
2.36621
5.307609
Quantify the following:
Importance of the phenomenon relative to money (1) ? 2
Difficulty of substituting other phenomena (.25) ? .25
Common ratio for an exponentially diminishing utility (.5) ? .5
The price and stock at saturation are 5.307609 and 2.36621
Aggregate utility is maximized near 5.307036 only.
There are 0 relative maximas greater than 5.710496
Press any key to continue

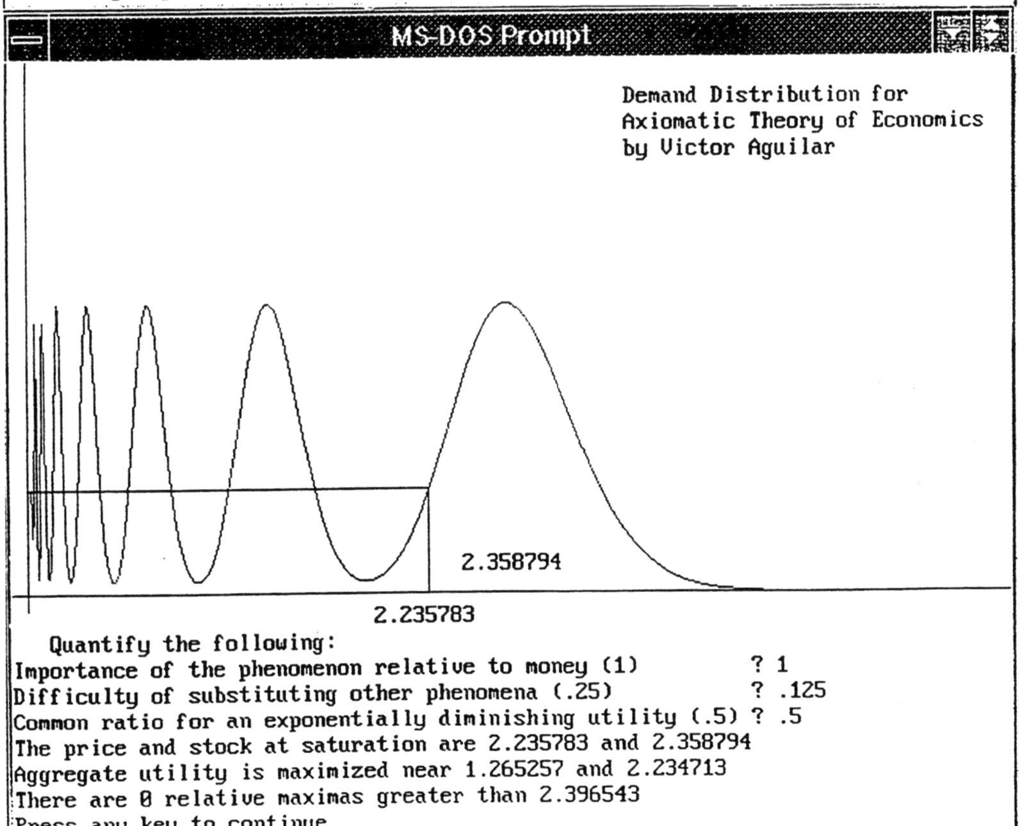
MS-DOS Prompt
Demand Distribution for
Axiomatic Theory of Economics
by Victor Aguilar
2.358794
2.235783
Quantify the following:
Importance of the phenomenon relative to money (1) ? 1
Difficulty of substituting other phenomena (.25) ? .125
Common ratio for an exponentially diminishing utility (.5) ? .5
The price and stock at saturation are 2.235783 and 2.358794
Aggregate utility is maximized near 1.265257 and 2.234713
There are 0 relative maximas greater than 2.396543
Press any key to continue

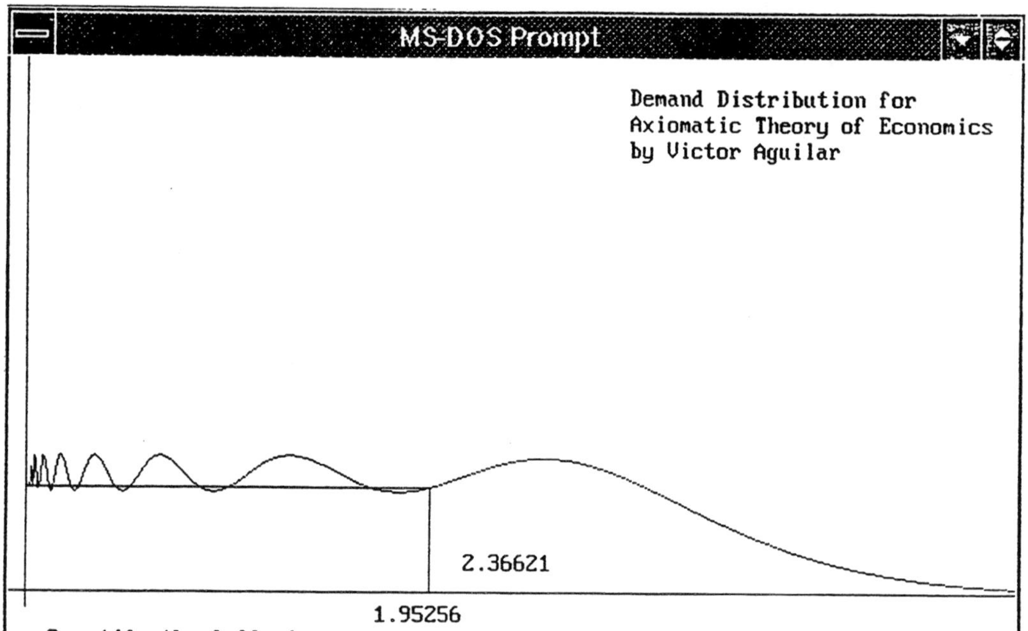


```
   Quantify the following:
Importance of the phenomenon relative to money (1)          ? 1
Difficulty of substituting other phenomena (.25)            ? .25
Common ratio for an exponentially diminishing utility (.5) ? .5
The price and stock at saturation are 1.95256 and 2.36621
Aggregate utility is maximized near 1.95235 only.
There are 0 relative maximas greater than 2.100774
Press any key to continue
```

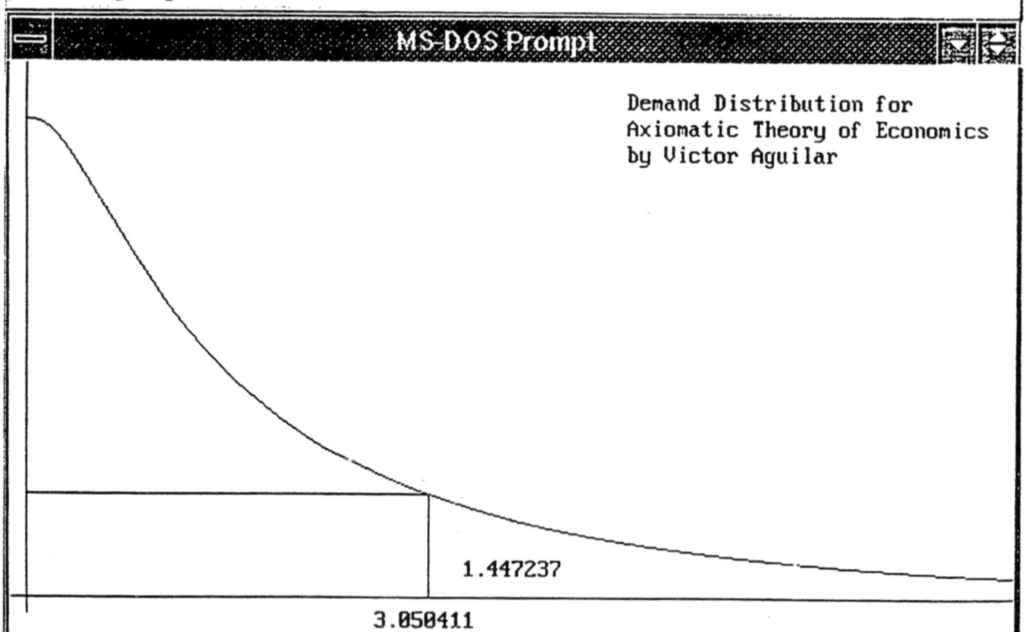


```
   Quantify the following:
Importance of the phenomenon relative to money (1)          ? 1
Difficulty of substituting other phenomena (.25)            ? 1
Common ratio for an exponentially diminishing utility (.5) ? .5
The price and stock at saturation are 3.050411 and 1.447237
Aggregate utility is maximized near 3.050369 only.
There is 1 relative maxima greater than .6608598
Press any key to continue
```

Appendix C: Software Manual for ECONOMIC CONSTRUCTION SET

I

Overview of the Software

The purpose of this software is to allow the user to define his own economic theory (within the framework of IS-LM analysis) and to test the accuracy of his model by comparing his predictions with historical data. It is not a game in which one attempts to maximize real national income by the judicious use of government spending or the money supply. **Economic Construction Set** asks the user for a descriptive, not a prescriptive, theory. It provides the user with an exercise in positive economics and does not judge his theory by any normative criteria. To define his theory, the user inputs mathematical formulas describing causal relations between the measured states of the economy and the levels of certain types of activity. If his formulas accurately reflect how people engaged in those activities respond to the state of the economy, the user is able to make predictions which mirror how the economy actually responded in various historical settings.

The state of the economy is described by three variables: national income, Y; interest charged on money, R; and the average price level, P. National income is how many billions of dollars change hands in a year, as reported to the Internal Revenue Service. The interest rate is the Federal Funds Rate. While the terms of these loans are very short, as little as a few hours, the rate is normalized to the ratio between the amount of money that would be owed after one year and the amount of money loaned, compounded yearly. The price level at a given time is the ratio between the weighted average of prices at that time and their weighted average at a time in the past which was chosen as the base year.

There are four historical variables which change over time: government spending, g; the supply of money, m; the stock of money, M; and the tax rate, r. Government spending is how many billions of dollars the government spends in a year. The supply of money is how many billions of dollars are "printed" in a year by the central bank. The process is a little more complicated than just printing money, but the details need not concern users of **Economic Construction Set**. The stock of money is how many billions of dollars have been printed to date. Economic textbooks often confuse the stock and supply of money, so it is important for the user to keep in mind that these are distinct quantities. The tax rate is the ratio between the Internal Revenue Service's receipts and national income.

This data is published by the federal government every quarter and **Economic Construction Set** interpolates it with a natural cubic spline so that the temporal scale can be graduated as finely as the user chooses. The user is not allowed to take steps longer than one quarter (three months), however, lest Y, R, and P change so much that **Economic Construction Set** cannot find their new values. In general, nothing is gained by taking steps shorter than one month. Regardless of the time increment, variables denoting a flow of money are normalized to billions of dollars per year. If the user cites the derivative of a historical variable when defining C, I, X, D, or A, **Economic Construction Set** will use the slope of the spline approximation to that historical variable. If the user cites the derivative of an output variable (Y, R, or P) when defining C, I, X, D, or A, **Economic Construction Set** will use the difference between that variable and its last calculated value, divided by the time interval. Since the last calculated value reflects the accumulated inaccuracies of the user's economic model and the current value is still unknown, the user should try to avoid citing derivatives of Y, R, or P, as they may cause small inaccuracies in the model to grow from one time period to the next. Derivatives of g, m, and r are as accurate as the historical data on g, m, and r. The effect of inherent (measuring) errors in historical data is discussed in Section IV.

All of the functions in this model are of seven variables: Y, R, P, g, m, r, and t with t representing time. Denote the unknown variables by **y** = [Y , R , P] and the known ones by **x** = [g , m , r , t]. **y** is the vector of variables for which the model is solved and **x** is the vector of variables which are taken from historical data. It is well known that to find isolated solutions one needs as many independent equations as one has variables in **y**. It is standard practice among mainstream economists to use the income identity, Y = C + I + g + X, the identity of stock and demand for real money holdings, D = M/P, and the existence of a price adjustment mechanism, $P = \prod A$. Let $\phi(\mathbf{x},\mathbf{y})$ be the vector of these three equations, each with everything brought to one side of the equals sign. For any given time and its associated **x**, the predicted **y** is the one such that $\phi(\mathbf{x},\mathbf{y}) = 0$. This is called the solution of the system and is compared to the historical values of Y, R, and P to judge the accuracy of the user's model. The program is called a "construction set" because, within the framework of the above-mentioned equations, there are five functions whose form determines how the model behaves, and these are left entirely up to the user to define. These five functions are: consumption, C; investment, I; net exports, X; demand for real money, D; and inflation (the price adjustment mechanism), A. Because the user has complete freedom in defining each of these functions in terms of Y, R, P, g, m, r, and t, he is under no restriction to act as a Keynesian, a Monetarist, or to support any other school of thought, but may define his own, as long as he believes in supply and demand.

No other economic software can justly be called a "construction set" because none of them allow the user to input his own functions. Instead, they have preset functions which reflect the author's own biases. Some of them allow the user to adjust certain numerical weights, but under no circumstance do they allow him to define his own theory of economics. At best, such programs can be compared to the "Hammurabi" type games in which a social/economic system is defined by the author and then the user is asked to operate in it, perhaps adjusting g, m and r, but accepting the defined system whether he finds it realistic or not. The principal reason that such programs prevent the user from inputing his own functions is that they cannot differentiate arbitrary, user-inputed functions symbolically, so they use predefined functions which the author differentiates on paper before programming the software. As will be shown in section III, in order to solve the system of equations, each coordinate function of $\phi(\mathbf{x},\mathbf{y})$ is differentiated with respect to all of the variables in both **x** and **y**; that is, 21 symbolic derivatives are calculated at run time. Without this ability, economic software is reduced to either a simple Hammurabi-type game with predefined functions, or it takes the entirely unrealistic position that all of the functions are linear. The former, while an enjoyable game, teaches nothing, for operating within another's theoretical system is not a creative activity. The latter, while it does allow the user to adjust the weights of each linear function, is so simplistic that the user's predictions are rarely close to the historical data. **Economic Construction Set** has the ability to symbolically differentiate arbitrary, user-inputed functions, so it makes no restrictions on the user. He may define complicated compositions of transcendental functions which are very difficult to differentiate by hand but which are handled easily by **Economic Construction Set**. There is no necessity for the artificial systems of functions so often found in textbooks where the derivatives all conveniently cancel out. Such functions do not reflect the author's economic insight but simply his wish to save the students from difficult algebra.

In **The Death of Economics**, 1994, Ormerod sharply criticizes mainstream economics for using only linear models of the economy and presents some data to demonstrate that certain relations (e.g. the Phillip's curve) are linear at times but have also made abrupt shifts. The abrupt shifts cannot be modeled with linear equations. In the past, economists simply started over with a new linear model when it became apparent that the old one was no longer working. However, by using **Economic Construction Set**, which has the ability to symbolically differentiate arbitrary, user-inputed functions, economists can address abrupt shifts in the economy. Whether or not this will save mainstream economics from its predicted death remains to be seen, though the use of non-linear equations is a big step forward. (It should be noted that Ormerod is not addressing IS-LM analysis but rather his own system of two non-linear equations in two unknowns, e, employment as a proportion of the labor force, and w, the share of wages in national output.)

In **Axiomatic Theory of Economics** I wrote "Business cycles take place over time, but mainstream economists insist on taking a snapshot of the economy and then analyzing it without any reference to the process that it is a part of." However, since then it has come to my attention that mainstream economists do have an explanation of business cycles. It is called *Nonlinear Business Cycle Theory* and, as the name implies, it can be modeled with **Economic Construction Set**. Paul Krugman writes[21] "I may be the only economist in my generation who has even heard of it" so I would suggest that interested readers consult Dr. Krugman for details. Typical university students will not run into this theory and are safe in assuming that there is no explanation of business cycles that is compatible with their studies. Mainsteam economist's primary contribution to business cycle theory is lots of advice (More government spending!) on how to get out of recessions when they do occur. Blame for the cycle itself is vaguely assigned to "animal spirits", "exogenous shocks", "irrational exuberance" or the like.

Section II of this manual discusses the five functions which the user inputs. It provides a number of helpful guidelines and should be read carefully. A familiarity with high school algebra is sufficient for inputing the five user-defined functions. Section III explains how the solution is found by **Economic Construction Set** and is of more mathematical interest. This section requires some knowledge of multivariable calculus and is not necessary for operating **Economic Construction Set.** Section IV addresses the issue of accuracy in the numerical calculations and also requires some knowledge of multivariable calculus. The user need not follow all of the math in this section but he should be aware of the existence of error and understand that it is inherent in the historical data and is not a failure of the software. Section V discusses the multiplier, which is defined in Section III, and discusses Keynesian and Monetarist policy recommendations.

The manual for **Economic Construction Set** can be read at two levels: At the operational level one should skim the more mathematical parts of sections III, IV and V and define C, I, X, D, and A without concerning oneself too much with how Y, R, and P are found by the computer so that the identities $Y = C + I + g + X$, $M/P = D$, and $P = \prod(A)$ are satisfied. At the research level, one should carefully read sections III, IV and V. These sections discuss a much more general method for solving simultaneous equations than is given to doctoral candidates in economics who have no convenient means of calculating symbolic derivatives. At that level they employ a trial and error approach to differentiating the several functions in the hopes that cancellations will occur. In academic situations, cancellations do occur which allow them to evaluate fewer derivatives than **Economic Construction Set** evaluates. To do research in macroeconomics, however, economists need **Economic Construction Set** because they can no longer count on all the algebra working out conveniently.

If one skims the more mathematical parts of sections III, IV and V, **Economic Construction Set** may be used at the high school level, though probably for an honors course. No previous study in economics is required; this manual contains a complete (though concise) explanation of modern, mainstream macroeconomics and is all the information necessary for operation of the software. The user is asked to input mathematical functions, however, and is expected to have a working knowledge of high school level algebra. Furthermore, the user is required to be creative in a very real sense. **Economic Construction Set** is not a tutorial which only asks questions for which it already has an answer. Neither does it frustrate the user by asking him to adjust numerical constants when he is aware that they have already been fixed as accurately as possible with methods unavailable to the user, such as linear regression. **Economic Construction Set** asks the user for functions (not constants) describing certain economic activities which he can successfully model only by acquiring insight into the causes of those activities. When fixing any numerical constants in his functions, the user has access to exactly the same multiple linear regression techniques that would be used at any level of research. While the user has complete freedom in describing consumption, investment, net exports, the demand for real money, and the inflation rate, he is not expected to build an economic theory from the ground up, which is too much to expect of non-professionals. The income identity, the identity of stock and demand for real money, and the existence of a price adjustment mechanism provide a framework for the user on which to build his theory, but they also restrain him from attacking the foundations of mainstream economic thought, which is inappropriate at the high school level. More sophisticated users with some mathematical background can profit from reading sections III, IV and V, which discuss the internal workings of **Economic Construction Set** and how to make economic forecasts. When studied at that level, **Economic Construction Set** becomes a useful tool for doctoral level research, especially if one takes advantage of the fact that we (Aguilar Software Company) are a software development company and can customize our software to fit the individual's needs.

II

The Framework of the Model

As noted above, $\phi(\mathbf{x},\mathbf{y})$ has three coordinate functions, each of which represents a fundamental identity which is satisfied at all times. As g, m, r, and t change over time, Y, R, and P adjust so that $\phi(\mathbf{x},\mathbf{y}) = 0$. Because this equality is an identity, it cannot fail to be true even for an instant because of how Y, R, and P are defined. For this identity not to be true is for Y, R, and P to have changed their definitions. While economists may disagree among themselves about how the user should input his functions, no mainstream economist can disagree with the three identities represented by the three coordinate functions of $\phi(\mathbf{x},\mathbf{y})$. All mainstream economists agree on the first identity, that national income is defined as the sum of consumption, investment, government spending, and net exports. Similarly, every mainstream textbook identifies the stock and demand for real money holdings (they say "supply" but in the case of money holdings they mean stock) and they all affirm the existence of a price adjustment mechanism. Thus, everyone must agree that the three identities represent mainstream thought.

Within these three identities are five functions which the user inputs and which define his theoretical position and determine the accuracy of his predictions. It is important for the user to realize that these may be input *ad libitum*, without any arbitrary restrictions imposed by weaknesses in the software. It is also important for him to realize, however, that some thought has been given to the problem of defining these five functions. At least in the beginning he should keep in mind certain restrictions economists usually impose on themselves when defining their theories. The purpose of this section is to discuss such guidelines for defining one's economic theory. In general, adherence to these guidelines will improve the accuracy of one's model. This is a construction set, however, and the user should not expect any recipes for success.

All of the economic variables discussed in this section are nominal, that is, unadjusted for inflation. Because **Economic Construction Set** is three dimensional, P is one of the variables predicted and one can get the real value of any variable simply by dividing it by P. This conflicts with some textbooks which define all of their variables as real, that is, as having already been adjusted for inflation. These textbooks include a price adjustment mechanism only as an afterthought and continue to use a two-dimensional system, which is called IS-LM analysis, simply mentioning that the level sets of the first two coordinate functions "shift" due to changes in the price level. They do not redefine their variables to be nominal even after introducing a price adjustment mechanism. However, using a two-dimensional system without a price adjustment mechanism can only be justified by the assertion that changes in the price level are inconsequential. But if that is true, why is any distinction between real and nominal values made at all during IS-LM analysis? When P is included among the unknown variables to be predicted, the other variables cannot have already been adjusted for changes in the price level because, by assumption, those changes are unknown until after the system of equations is solved. Thus, any model which predicts the price level can only do so with nominal values for the known variables. Degenerate systems which do not predict the price level assume that inflation is inconsequential and, therefore, cannot justify the need for real variables. **Economic Construction Set** has all three dimensions and uses only nominal values so that it may predict the price level. It may also model economies in which changes in the price level are considered inconsequential if one wishes to simulate IS-LM analysis.

The income identity asserts that national income is composed of consumption, investment, government spending, and net exports. Thus, $Y = C + I + g + X$, or equivalently, the first coordinate function of $\phi(\mathbf{x},\mathbf{y})$, $C + I + g + X - Y$, is zero at all times. g is fixed historically and Y is an unknown variable, but C, I, and X are defined by the user to be functions of Y, R, P, g, m, r, and t, though not necessarily of all of them. The following three paragraphs provide guidelines for defining C, I, and X. After that is a paragraph discussing the demand for real money, D, and then several paragraphs are devoted to inflation, A. This rounds out the five user-defined functions, C, I, X, D, and A.

Consumption Consumption and savings together exhaust all possible uses of disposable income, (1 - r)Y. Thus, one expects consumption, C, to be negatively related to the interest rate, R. This is called the substitution effect of interest rates on consumption because people substitute savings for consumption. It may be more realistic to consider the real interest rate, R/A, however, since a high return does not mean as much if prices have increased by the time one is able to collect one's money. Furthermore, the interest rate offered by the banks on savings accounts means more to people with high disposable incomes than those with low disposable incomes because the former group has more money to play with. Those with low disposable incomes consume most of their earnings just to survive and do not have much left over to save. Thus, consumption is a positive function of disposable income. When interest rates rise, people's income from their assets increases, which make them want to consume more. This is called the income effect of interest rates on consumption. Most mainstream economists believe that the income and the substitution effect of interest rates on consumption cancel each other out. Thus, people consume a fixed proportion of their disposable income regardless of the interest rate, that is, C = b(1 - r)Y with b the marginal propensity to consume. The derivative of C with respect to Y is b, so it can be determined either by linear regression on C and (1 - r)Y values over time or by averaging the ratio of changes in C with changes in (1 - r)Y over time. The latter method usually results in a smaller coefficient because consumption is more stable than income. This is because most people do not change their consumption patterns much in response to temporary changes in their income. Consumption does become more conservative if people expect a lengthy depression, however, so one may consider making C a negative function of the deficit relative to national income, g/Y - r, because government waste is the primary cause of depressions. Furthermore, consumer goods may be broken up into durables, non-durables, and services, so one may define a consumption function for each of these and then add them together. Consumer durables, such as automobiles, are often financed and so they depend on the interest rate more than the other categories. They are also more dependent on temporary income swings because they can be foregone more easily than non-durables such as food. In fact, they behave very much like investments, which are considered in the next paragraph.

Investment The proportion of investment to national income is principally a function of the interest rate because it reflects spending on goods which bring their owner a return. If that return exceeds the interest he must pay on the loan needed to buy them (or, if he paid cash, the foregone interest he could have earned had he loaned the money out himself), the purchase is made. The lower the interest rate, the easier it is for purchases to qualify as good investments, so investment, I, is negatively related to the interest rate, R. Denote one's profit (ratio of income to expenses) from an investment by p so, without taxes, a good investment is one in which $R < p$. Because of competition, investments are made until profits are reduced to the interest rate, which is why I is a function of R and not p. With taxes, one's net profit is one's gross profit minus the product of the tax rate and the difference between gross profit and unity (breaking even), that is, $p - r(p - 1)$. Thus, a good investment is one in which $(R - r)/(1 - r) < p$. It may be more realistic to multiply $(R - r)/(1 - r)$ by A, however, because investment income is denominated in depreciating currency while the loan one received to make that investment is payable in a more stable foreign currency or in commodities. For instance, one may owe a certain quantity of corn every year if one borrowed to buy a farm. Loans denominated in things other than dollars are rare in America because, historically, the inflation rate has not been a serious problem, but the principle still holds. Furthermore, investment goods may be broken up into non-residential fixed investments, residential fixed investments, and inventory investments, so one may define an investment function for each of these and then add them together. Residential investments are very sensitive to the interest rate because houses depreciate so little that their cost is dominated by interest payments. Inventory investments are insensitive to the interest rate. In fact, they are almost a fixed percentage of national income because a given flow of goods requires a certain amount "in the pipeline."

Net Exports Recall that consumption expenditures are primarily a positive function of income. However, many of the consumption goods available are manufactured in other countries so, when national income rises, part of the increased demand for goods leak into other economies from which we import. Net exports are defined as exports minus imports so they are negatively related to income or, more accurately, to disposable real income. Furthermore, if an excess of goods flows in or out of a country (they have a trade imbalance), the difference is made up with financial instruments. In other words, to buy goods worth more than those one sells, one must borrow. Thus, net exports are negatively related to the interest rate because to offer a high interest rate indicates that one is anxious to borrow. Another way to look at this is that a high interest rate makes financial instruments denominated in the domestic currency attractive to foreigners. To get the domestic currency they need for buying our financial instruments, they sell us their goods. Yet another way to look at this is that when domestic currency becomes more attractive to foreigners, they bid up its price (the exchange rate), and a high exchange rate discourages net exports. By a "high" interest rate, I mean high relative to foreign interest rates. But their Y, R, and P are reciprocally related to ours (they both depend on each other), so it is impossible to completely model net exports by considering only conditions in the United States. However, the United States is large enough that it is a reasonable approximation to consider only domestic data. To model the world economy requires 3n nonlinear equations of 6n + 1 variables for 3n unknown variables, where n is the number of independent countries (with their own central bank) in the world, about 30. Fortunately, this is a fairly sparse system, since most countries have only a few principle trading partners. Nevertheless, solving such a system is a bigger project than the educational software market would justify. **Economic Construction Set** as it stands is already somewhat of an overkill for academia and we would not consider setting up and solving a function $\phi\Re^{181}\rightarrow\Re^{90}$ unless someone (like a national government) paid us to do so; the academic market does not demand such sophisticated software.

Demand for Real Money The identity of stock and demand for real money holdings asserts that this market always clears, that is, M/P = D with D the user inputed demand function for real money holdings. Equivalently, the second coordinate function of $\phi(\mathbf{x},\mathbf{y})$, D - M/P, is zero at all times. The demand for real money holdings is positively related to income because one needs to hold currency for the day-to-day transactions one makes, and the higher one's income, the more transactions one makes every day. Mainstream economics asserts that the demand for real money holdings is also negatively related to the interest rate because money that one is holding in currency is not earning any interest and, if the interest rate is high, one conserves on one's money holdings to put more in the bank. Concomitantly, if the interest rate is low, one holds large quantities of currency in speculation of the time when interest rates rise again so that, when they do rise, one has money on hand to buy financial instruments. However, economists increasingly assert that most people hold as little of their wealth in money as is convenient regardless of the interest rate. Also, when the interest rate is low, nobody holds money in speculation of it rising because they know that the government can hold short-term interest rates down indefinately by issuing money. The user can make D a function of R at his own discretion.

Price Adjustment Mechanism The existence of a price adjustment mechanism simply asserts that there is a causal relation between the state of the economy and the inflation rate, that is, the temporal change in the price level, P. Let time, t, be defined so that it is zero at the beginning of the historical period in question and the price level, P, be defined to be unity. All later price levels are expressed relative to the original value of the currency in that historical period. Hence, P = Π(A) with A the user-inputed price adjustment mechanism. Equivalently, the third coordinate function of $\phi(\mathbf{x},\mathbf{y})$, Π(A) - P, is zero at all times. This permits a consistent adjustment for inflation during the historical period in question. Choosing an arbitrary base year like

1982 does not work because there is nothing significant about 1982 and changing that alters the solution to the model even with the same price adjustment mechanism.

There is no general consensus regarding the mechanism by which prices adjust, and the form of this function does the most toward defining the user's theoretical stance. Because of this disagreement and because many textbooks ignore price adjustment and simply solve the first two coordinate functions for Y and R (what is traditionally referred to as IS-LM analysis), much of what follows regarding the third coordinate function reflects my own opinion. Some economists may actually dispute the existence of a third coordinate function because their own education never got past IS-LM analysis. Denoting the real interest rate by R/A and relating A and R as discussed below can only work if both numbers are ratios around the multiplicative identity, unity. All mainstream economists denote interest by a number around the additive identity, zero. The motivation for this is their interest in using naive multiple linear regression (By "naive" I mean without any transformation of variables.) to define functions such as M/P = kY - hR which treat both Y and R as flows. However, R is not a flow, so the units in the above equation are wrong. In classical theory, M/P = kY was called the quantity theory of money. Since the units of Y are dollars per year and the units of M are dollars (P is dimensionless), the units of k must have been years. They were, in fact, the average number of years people held onto a dollar before spending it. But when the Keynesians added the hR term, the units no longer worked out. A similar argument can be made against net exports being a linear combination of Y and R.

All mainstream economists assert that the real interest rate is the difference between the nominal interest rate and the inflation rate. For example, if a savings account yields 10% per year and prices are increasing by 6% every year, they assert that one earns only 4% in real terms. A simple example demonstrates this to be wrong: Consider a man with \$100 who is interested in purchasing \$1.00 items. Immediately, he can buy 100 items. After a year of saving, he has \$110 but the items he wants each cost \$1.06. He can buy 110/1.06 = 103.77 items. By foregoing 100 of these items for a year, he can earn an extra 3.77

items. Obviously, the real interest rate is R/A = 1.10/1.06 = 1.0377, not 1.04 as mainstream economists assert. Users who prefer the traditional measure of the interest rate despite this problem or who are testing their textbook's theory may simply use R - A rather than R/A for the real interest rate. Thus, the mistake of subtracting the inflation rate from the interest rate is easily excised from mainstream economics or left standing, as the user prefers.

A and R are both the ratio of something with the same thing measured a year before. All the other variables (except the tax rate, r, and the price level, P, which is the running product of A) are flows over time, that is, the difference between something and the same thing measured a year before. Thus, one expects the inflation rate and the interest rate to be related. This is indeed true because inflation and the interest rate are both directly affected by the money supply, the former because people buy things with the new money and the latter because banks use the new money to make loans. Hence, the relation between A and R should be of the form $A = R^{f(m)}$ with f defined by the user. f is a positive function of m; when new money is issued it drives the interest rate down and the inflation rate up. This assertion is also supported by the fact that interest rates and price adjustments are routinely played against each other when negotiating for expensive items. For example, a buyer who wishes to purchase a car is always given a choice between a rebate and a low-interest loan. Most economists agree that changes in the price level are positively related to the money supply, m, though defining a quick adjustment mechanism qualifies one as a classical thinker while a sticky adjustment mechanism indicates a follower of Keynes. The money supply, m, has more influence on the state of the economy than any other measured parameter, and a great deal of effort is made in predicting it. Unfortunately, changes in the money supply are unannounced and, in fact, concealed for as long as possible so that the government may spend the new money before people react by raising prices. Also, the money supply has more volatility because it is controlled by one man, the Federal Reserve Chairman, while government spending and the tax rate are controlled by lethargic bureaucracies. However, the principal reason for issuing money is to pay for the government debt, G - rY, which is the amount not paid for with taxes. If government debt is high relative to national income, one can expect the price level to rise in anticipation of a new issue of money. Another predictor of future supplies of money from the federal government is an upcoming election. This could be modeled by the function $\cos\left(\frac{\pi t}{2} + \theta\right)$ with the phase angle, θ, equal to the time until the next election multiplied by $\frac{\pi}{2}$. Thus, if one defines $A = R^{f\left(m\,,\, \frac{G}{Y\text{-}r}\,,\, \cos\left(\frac{\pi t}{2} + \theta\right)\right)}$ one has included both the direct effect of the money supply on prices as well as the indirect effect of people anticipating future money supplies.

One difficulty with defining the price adjustment mechanism, A, is that, after inflation has become a problem, people automatically increase their prices to stay ahead of their depreciating currency. This gives prices a life of their own that does not respond to measured economic variables in a predictable way. Another difficulty is that, in response to inflation, the government often imposes regulatory or legal obstacles to loaning money to the public, so businesses see very little new money even during periods of low interest rates. This is called a credit crunch. Since government policy, both in issuing money and imposing a credit crunch, is unpredictable and changes with the politics of the day, people's automatic price increases (based on what they see as a government policy to devalue the currency) and the availability of loans cannot be modeled effectively. This is particularly true in countries like the United States where the government's monetary policy is determined by one person or by a small group of people who are not obliged to explain their actions or to respond to elected officials.

In mainstream economics, the price adjustment rate is a function of national income, Y, relative to its "potential." They assert that when national income is greater than potential, demand can no longer be met by increased output because the physical limits on production have been reached, so prices increase instead. When the economy is experiencing a depression, they assert that price adjustments are slow because there is excess capacity in the production system. Many economists take a position between the classical and Keynesian positions by asserting that A is flat at income levels below potential and bends upward at income levels above potential. This is equivalent to considering the classical model as being appropriate during full output and the Keynesian model as being appropriate during depressions. The AS-AD graph seen in mainstream textbooks is used to explain this relation.

Whether or not this relation is true, it is impractical unless one can attach some meaning to the very nebulous concept of potential national income. Okun's Law, which was derived empirically, expresses a (linear) positive relation between employment relative to its potential and national income relative to its potential. Without normalizing employment and national income to their potential values, the identification of employment and national income is usually referred to as the production function. The Phillip's curve, also empirical, expresses a negative relation (a trade-off) between inflation and employment relative to its potential. Putting the two together, the Phillip-Okun's curve describes a positive relation between the inflation rate, A, and national income relative to its potential. Employment need not appear in the model because it can be derived from national income, which the model predicts. In the historical period considered by **Economic Construction Set**, there is a 100% correlation between national income and employment within their measured accuracy. This 100% correlation demonstrates the linearity of the production function but it does not help one define the Phillip-Okun's curve, which addresses potential employment and potential national income.

Neither the capital stock nor the employment level appear in **Economic Construction Set.** Mainstream economics is entirely demand-side and the *ad hoc* introduction of capital and employment is a fundamental departure from its philosophical underpinnings. Also, the capital stock is an absolute quantity while all the other variables in **Economic Construction Set** are flows, so it is impossible to make the units work out. More importantly, it is virtually impossible to measure either the stock of capital or the flow of labor with even one or two significant digits of accuracy, and it would be a disservice to the user to provide data on capital and employment as though it were meaningful. In fact, it is well known that the result of the Cambridge Controversies (sometimes referred to as the Capital Controversies) was that the aggregate production function, Y = f(N,K), is meaningless. Neither labor, N, nor capital, K, can be measured empirically and, more to the point, their aggregation is logically inconsistent. Furthermore, there are logical problems with the existence of a function mapping aggregate labor and capital into national income.

Nevertheless, the Phillip-Okun's curve is mainstream while $A = R^{f\left(m,\ \frac{G}{Y-r},\ \cos\left(\frac{\pi t}{2}+\theta\right)\right)}$ is of my own invention. If one would rather use the Phillip-Okun's curve to predict inflation, one needs to do time-series analysis on capital and employment to determine potential national income. For this, one needs census data on the size of the population and a method to quantify the skills possessed by people to determine potential employment. Quantifying skill levels is the most important part of such an analysis because skilled people performing manual labor have more in common with the unemployed than with the employed. Researchers with such an analysis in mind should contact us (Aguilar Software Company) at the address given in the back of the manual to have their software customized to suit their needs.

Degenerate Systems Many economists hold the price level fixed; that is, they define A = 1. Users who are testing their professor's theory with **Economic Construction Set** should keep this in mind, for it is often tacitly assumed. In general, when one sees the phrase, "IS-LM analysis", it refers to setting A = 1 and solving only the first two coordinate functions, the level sets of which are called the IS and LM curves, for Y and R. In lower-division courses, professors often assume not only that prices are fixed but also that the interest rate is fixed. At that level, only the income identity is defined so investment is autonomously determined by "animal spirits" and net exports are either ignored or made a function of income alone. The only unknown variable is Y. The graph of the two sides of the income identity (Y and C + I + g + X) are referred to as the "Keynesian Cross". Users who are testing their professor's theory in this case cannot simply omit R from all the user-defined functions because that produces a singular system of equations, resulting in the nil determinant mentioned in section III. To test such a theory, the user need only solve the first coordinate function of $\phi(\mathbf{x,y})$, C + I + g + X - Y. Usually, C is a linear function of Y while I, g, and X are constants, so the equation can be solved for Y algebraically. At the worst, it can be solved for Y using the one-dimensional Newton's method algorithm available on any programmable calculator. It should be obvious that such a theory has no correlation with historical data. To use sophisticated software such as **Economic Construction Set** to demonstrate its limitations would be an overkill.

Summary The preceding discussion exhausts the list of user-defined functions, so this paragraph will summarize the meaning of all the variables and functions defined by **Economic Construction Set**. There are three variables which describe the state of the economy: national income, Y, a flow; the interest rate, R, a ratio; and the price level, P, a ratio. There are five historical variables which change at discrete time intervals: government spending, g, a flow; the supply of money, m, a flow; the stock of money, M, which is measured in dollars; the tax rate, r, a ratio; and time, t, which is measured in years. There are five user-defined functions: consumption, C, a flow; investment, I, a flow; net exports, X, a flow; demand for real money holdings, D, which is measured in dollars; and inflation, A, a ratio. All of the variables and functions denoted as flows are denominated in billions of dollars per year. Most mainstream economics textbooks confuse the words "stock" and "supply", especially when discussing money, and the user should be aware that this manual is consistent with "supply" denoting a flow and "stock" denoting the accumulation of the corresponding flow over a time period.

Flows may be added to each other to get other flows. Multiplying two flows together is meaningless. Stocks are flows multiplied by variables which measure the duration of an event. All of the variables and functions denoted as ratios are dimensionless because their units cancel out when the division is made. Ratios may be multiplied with each other to get other ratios. Adding ratios, either to each other or to flows, is meaningless. Ratios may be multiplied with flows to weight the significance of that flow in a function. A word to the wise: **Economic Construction Set** measures one's insight into economic theory and the least insightful thing one can do is to confuse one's units of measurement. Physicists employ seven basic units (meter, kilogram, second, ampere, Kelvin, mole, and candela) and are in the habit of keeping track of them while macroeconomists employ only two, dollars and years, but often define functions like $M/P = kY - hR$, which are meaningless. Be faithful to the fundamentals; do not confuse one's units of measurement.

The preceding paragraphs provide guidelines for defining the five functions which represent the user's theory of economics. These guidelines are by no means prescriptive and, in fact, no general consensus exists among professional economists regarding the definitions of the user-inputed functions. While suggestions are made as to whether each of these functions should be positively or negatively related to this or that variable, it is up to the user to define his own functions. If his definitions prove insightful, his model will be highly correlated with the historical data. If one is able to define a model which accurately describes how past situations worked out, one can predict the future effects of a government reaction function, that is, a predictable way in which government spending, the money supply and the tax rate respond to economic indicators. **Economic Construction Set** measures one's understanding of the economic impact of the government's actions but does not predict what actions the government will take. To predict the future requires removing g, m, and r from **x** and making them user-defined functions. These functions are more difficult to define than C, I, X, D, and A, however, because they are determined politically and the current state of the economy has only an indirect effect on politics. Section V discusses how to find a steady money supply, m, which optimizes a measure of the success of an administration, assuming that g and r react to the economy in a predictable way.

Hopefully, users who are studying macroeconomics in school can also use this software to test their professor's theory, thus inserting a bit of accountability into academia which has previously been lacking. Without any convenient method of verifying textbook theories, college students are easily and often deceived. As noted above, all mainstream economists accept the three identities upon which this software is based. Thus, without breaking with mainstream economics, one must accept **Economic Construction Set**'s correlation coefficient as a measure of one's theory. Even at the undergraduate level where one is given degenerate systems such as the Keynesian Cross or IS-LM analysis, one still accepts the three identities upon which this software is based. One is simply accepting them to include (unrealistic) constant functions.

Users who are interested in acquiring a more thorough understanding of economic theory are referred to my book, **Axiomatic Theory of Economics**. [Note: "I" refers to myself, Victor Aguilar, while "we" refers to Aguilar Software Company. The opinions expressed in **Economic Construction Set** are not necessarily those of Aguilar Software Company, or any of its employees.] Unlike this software, which requires only a familiarity with high school algebra to operate, **Axiomatic Theory of Economics** prerequisites three semesters of calculus and a semester of probability. **Axiomatic Theory of Economics** deduces all of its results from axioms and makes no reference to statistics. It provides a new perspective, one based on stock rather than supply (flow), so the concept of wealth replaces that of income. It is not a textbook on mainstream economics. Users who have questions about the underpinnings of the mainstream model which **Economic Construction Set** uses should consult a standard textbook. While every effort has been made to make **Economic Construction Set** reflect mainstream economics, it is not the purpose of this manual to serve as a textbook. Neither is it the purpose of this manual to define my own theoretical position. **Axiomatic Theory of Economics** is the authoritative statement on my position.

Readers of **Axiomatic Theory of Economics** will see that I am not an adherent of mainstream economics and, indeed, some criticism of it has appeared in this manual. Nevertheless, every effort has been made to make this software representative of mainstream thought. It is doubtful that any professor who is employed at a university in the United States could disagree with the three fundamental identities represented by the three coordinate functions of ϕ. Researchers may not be interested in being purely mainstream, however, and some economists may differ in their view of mainstream economics. (To get grant money, at least the first two identities must be kept intact.) Any user who would like a similar program to solve a different system of equations, perhaps even one with four or more coordinate functions, may write to us at the address given in the back of the manual. For instance, many textbooks discuss the capital stock and the level of employment and some economists may wish to see these mentioned explicitly in the model. Also, there is a great deal of interest in applying statistical techniques to time-series of data and some economists may like to have more statistical tools available than just multiple linear regression. Perhaps most useful of all, particularly to speculators in the bond market, would be the inclusion of long-term interest rates. This would require estimating future depreciations of the currency, which would require more sophisticated forecasting techniques than come with **Economic Construction Set**. Swapping out coordinate functions and adding statistical routines is relatively simple and the user is encouraged to have his software customized to suit his own needs. Throughout the development of this software, we made it as general as possible so that the routines for solving the three simultaneous, non-linear equations could be easily extended to larger systems without any loss of accuracy. We are a software development company and will gladly accept orders for everything from alterations of existing software to writing complete programs. Standard programming rates apply. Know what you want done. I have my own theory of economics (see my book, **Axiomatic Theory of Economics**) and it is not my business to explain mainstream economics to its practitioners.

III

How the Solution is Found Mathematically

As noted in the preceding section, $\phi(\mathbf{x},\mathbf{y})$ has three coordinate functions, each of seven variables, so $\phi:\Re^7 \rightarrow \Re^3$, where $\Re^n$ denotes n dimensional Euclidean space. Denote the three coordinate functions with the letters F, G, and H so $\phi(\mathbf{x},\mathbf{y}) = \begin{bmatrix} F(\mathbf{x},\mathbf{y}) \\ G(\mathbf{x},\mathbf{y}) \\ H(\mathbf{x},\mathbf{y}) \end{bmatrix} = \begin{bmatrix} C+I+g+X-Y \\ D-M/P \\ \Pi(A)-P \end{bmatrix}$ At every instant in time $\mathbf{x}$ is given, so the question is: What value of $\mathbf{y}$ is such that $F(\mathbf{x},\mathbf{y}) = G(\mathbf{x},\mathbf{y}) = H(\mathbf{x},\mathbf{y}) = 0$, simultaneously? It is well known that, were ϕ real-valued and $\mathbf{y}$ contained only one variable, say, y (ϕ must have as many coordinate functions as there are unknown variables in $\mathbf{y}$), then Newton's method asserts that successive iterations of $y_{i+1} = y_i = \dfrac{\phi(\mathbf{x},y_i)}{\phi_y(\mathbf{x},y_i)}$ should converge to a y such that $\phi(\mathbf{x},y_i) = 0$. Convergence implies the existence of an N such that, for all $n > N$, $\|\phi(\mathbf{x},y_n)\|$ is less than a specified tolerance of error. The convergence is not guaranteed: $\phi(\mathbf{x},y_n)$ must be smooth and the initial guess, y_0, must be sufficiently close to the solution that the algorithm does not get lost. In the equation above, $\phi_y(\mathbf{x},y_i)$ denotes the derivative of ϕ with respect to y, evaluated at $\mathbf{x},y_i$. Similarly, in three dimensions, $\mathbf{y}_{i+1} = \mathbf{y}_i = \phi_{\mathbf{y}}(\mathbf{x},\mathbf{y}_i)^{-1}\phi\,(\mathbf{x},\mathbf{y}_i)$ where $\phi_{\mathbf{y}}(\mathbf{x},\mathbf{y}_i)$ denotes the derivative of ϕ with respect to $\mathbf{y}$ evaluated at $\mathbf{x},\mathbf{y}_i$. The exponent, -1, represents the multiplicative inverse of what precedes it. What precedes it, however, is a derivative with respect to a vector, not with respect to a single variable as before, so it is not a real-valued function which may be divided into unity to get its multiplicative inverse. Rather, the derivative of a vector-valued function of several variables is a matrix of functions, the rows corresponding to the coordinate functions and the columns to the several variables. Thus, the derivative has as many rows as the dimension of the range and as many columns as the dimension of the domain. It is a theorem of linear algebra that a multiplicative inverse can exist only for square matrices, and that is why there must be exactly as many variables in **y** as coordinate functions in ϕ.

$$\phi(\mathbf{x},\mathbf{y}) = \begin{bmatrix} F_Y(\mathbf{x},\mathbf{y}) & F_R(\mathbf{x},\mathbf{y}) & F_P(\mathbf{x},\mathbf{y}) \\ G_Y(\mathbf{x},\mathbf{y}) & G_R(\mathbf{x},\mathbf{y}) & G_P(\mathbf{x},\mathbf{y}) \\ H_Y(\mathbf{x},\mathbf{y}) & H_R(\mathbf{x},\mathbf{y}) & H_P(\mathbf{x},\mathbf{y}) \end{bmatrix}$$

It is well known from linear algebra that the multiplicative inverse is the adjoint matrix divided by the determinant. The adjoint matrix is the transpose of the matrix of cofactors, the cofactor of the $a_{i,j}$ element being $(-1)^{i+j}$ times the determinant of the matrix with its i'th row and j'th column removed. Notice that the division is defined only for matrices with non-zero determinants; in fact, this is equivalent to the requirement that F, G, and H be independent. This requirement cannot be avoided by calculating Newton's method differently because it is the basic definition and any other method is derived from it.

If one were solving the system of linear equations

$$\begin{bmatrix} F_Y(\mathbf{x},\mathbf{y}) & F_R(\mathbf{x},\mathbf{y}) & F_P(\mathbf{x},\mathbf{y}) \\ G_Y(\mathbf{x},\mathbf{y}) & G_R(\mathbf{x},\mathbf{y}) & G_P(\mathbf{x},\mathbf{y}) \\ H_Y(\mathbf{x},\mathbf{y}) & H_R(\mathbf{x},\mathbf{y}) & H_P(\mathbf{x},\mathbf{y}) \end{bmatrix} \begin{bmatrix} u \\ v \\ w \end{bmatrix} = \begin{bmatrix} F(\mathbf{x},\mathbf{y}) \\ G(\mathbf{x},\mathbf{y}) \\ H(\mathbf{x},\mathbf{y}) \end{bmatrix}$$

for [u , v , w], one could premultiply both sides by the inverse of the square matrix so [u , v , w] remained alone on the left hand side. Thus, calculation of the adjustment at each iteration of Newton's method, [u , v , w] is equivalent to solving a system of linear equations. Numerical analysis provides many techniques for accomplishing this which are faster than premultiplying the right-hand side by the inverse of the square matrix. In only three dimensions, however, execution time is not an overriding concern and the elements of the inverse matrix are needed for analyzing inherent errors, which are discussed in the next section.

Given an initial guess at **y**, Newton's method provides an iterative method of approaching the solution of $\phi(\mathbf{x},\mathbf{y}) = 0$ which may be halted as soon as the change in **y** is less than a specified tolerance of error. It remains to be shown how that initial guess is attained. Consider the function $\mathbf{y} = \psi(\mathbf{x})$, which returns the **y** such that $\phi(\mathbf{x},\mathbf{y}) = 0$ for a given value of **x**. $\mathbf{y} = \psi(\mathbf{x})$ defines a surface in seven dimensional space, but not necessarily one which is known explicitly. That would imply that F, G, and H could be algebraically manipulated so Y, R, and P were each functions of g, m, r, and t. In general, such algebraic manipulation is impossible or at least impractical, despite the fact that this is the standard method of graduate level economics textbooks, which always set up their functions so that the manipulation works. The derivative of $\psi(\mathbf{x})$ is written $\psi_{\mathbf{x}}(\mathbf{x},\mathrm{y})$. Both **x** and **y** may appear in the functional notation of ψ and $\psi_{\mathbf{x}}$. ψ is a function only of **x**, since only one **y** is associated with each value of **x**, but **Economic Construction Set** does not attempt to isolate **y** algebraically. (Example: y = 2x - y could be simplified to y = x, but if one does not make this simplification, one could write $y = \psi(x,y) = 2x - y$ and it would still be true that only one y is associated with each x.)

If the slope of the surface is known, at any point on it a vector tangent to that surface may be drawn whose projection onto **x** space represents changes in g, m, r, and t. Where this vector points is a good approximation of **y** after adjustment of **x**. In fact, it is the first degree Taylor approximation. A zero degree Taylor approximation is just the old **y** and a second degree Taylor approximation uses a vector bent in the shape of a parabola. It is well known that the formula for the first degree Taylor approximation to ψ around a point in the domain, **a**, is $\mathbf{y}_1 = \mathbf{y}_0 + \psi_{\mathbf{x}}(\mathbf{a},\mathbf{y}_0)(\mathbf{x} - \mathbf{a})$ where **a** is the old value of **x** and $\mathbf{y}_0 = \psi(\mathbf{a})$. $(\mathbf{x} - \mathbf{a})$ is the change in g, m, r, and t since the system of equations was solved last. Thus, to describe the method of arriving at an initial guess at **y**, it is sufficient to show how one finds $\psi_{\mathbf{x}}(\mathbf{a},\mathbf{y}_0)$, that is, how one finds the slope of the surface of solutions to $\phi(\mathbf{x},\mathbf{y}) = 0$ at the point $\mathbf{a},\mathbf{y}_0$.

Using the chain rule to differentiate $\phi(\mathbf{x},\mathbf{y}) = 0$ with respect to $\mathbf{x}$, one gets $\phi_{\mathbf{x}}(\mathbf{x},\psi(\mathbf{x},\mathbf{y})) + \phi_{\mathbf{y}}(\mathbf{x},\psi(\mathbf{x},\mathbf{y}))\psi(\mathbf{x},\mathbf{y}) = 0$ or equivalently, $\psi(\mathbf{x},\mathbf{y}) = -\phi_{\mathbf{y}}(\mathbf{x},\psi(\mathbf{x},\mathbf{y}))^{-1}\phi_{\mathbf{x}}(\mathbf{x},\psi(\mathbf{x},\mathbf{y}))$ which, evaluated at $\mathbf{a},\mathbf{y}_0$, is

$$-\phi_{\mathbf{y}}(\mathbf{a},\mathbf{y}_0)^{-1} = \begin{bmatrix} F_g(\mathbf{a},\mathbf{y}_0) & F_m(\mathbf{a},\mathbf{y}_0) & F_r(\mathbf{a},\mathbf{y}_0) & F_t(\mathbf{a},\mathbf{y}_0) \\ G_g(\mathbf{a},\mathbf{y}_0) & G_m(\mathbf{a},\mathbf{y}_0) & G_r(\mathbf{a},\mathbf{y}_0) & G_t(\mathbf{a},\mathbf{y}_0) \\ H_g(\mathbf{a},\mathbf{y}_0) & H_m(\mathbf{a},\mathbf{y}_0) & H_r(\mathbf{a},\mathbf{y}_0) & \ln(A)A^t(\mathbf{a},\mathbf{y}_0) \end{bmatrix}$$

This is a product of two matrices, the first of which was already calculated for use with Newton's method. Notice that $\phi_{\mathbf{y}}(\mathbf{x},\mathbf{y})$ has to be square so that its inverse is defined. $\phi_{\mathbf{x}}(\mathbf{x},\mathbf{y})$ must have three rows but need not have three columns. A 3×4 matrix premultiplied by a 3×3 matrix results in a 3×4 matrix which is exactly what is needed to transform $\mathbf{x} - \mathbf{a}$ from four dimensional space into three dimensional space so that it may be added to $\mathbf{y}_0$. As with Newton's method, the evaluation of $\psi_{\mathbf{x}}(\mathbf{x},\mathbf{y})$ is equivalent to solving a linear system of equations; in this case, one with four right-hand column vectors and as many solution vectors: $-\phi_{\mathbf{y}}(\mathbf{x},\mathbf{y})\psi\mathbf{x}(\mathbf{x},\mathbf{y}) = \phi_{\mathbf{x}}(\mathbf{x},\mathbf{y})$.

$\psi_{\mathbf{x}}(\mathbf{x},\mathbf{y})$, the slope of the vector tangent to the surface of solutions to $\phi(\mathbf{x},\mathbf{y}) = 0$ whose projection onto $\mathbf{x}$ space represents changes in g, m, r, and t, could be called the multiplier. The word "multiplier" is used at the lower-division undergraduate level where only the income identity is defined (so ϕ has only one coordinate function, the first), the only variable in $\mathbf{y}$ is Y, and the only variable in $\mathbf{x}$ is g. Here, one is solving $\phi(g,Y) = 0$ for Y using the standard one-dimensional Newton's method algorithm that everybody is familiar with from first semester calculus. $Y = \psi(g)$ is the set of solutions to $\phi(g,Y) = 0$. Like $\mathbf{y} = \psi(\mathbf{x})$ which defines a surface in seven-dimensional space, $Y = \psi(g)$ defines a curve in two-dimensional space, the g, Y plane. Thus, $\psi'(g)$ tells one how much national income increases in response to increases in government spending. It is called a multiplier because economists teaching at the lower-division level in college assert that increases in government spending cause national income to increase by a proportionally larger amount. If presidential advisors used $\psi'(g)$ as the basis for their policy recommendations, they would always advise an infinite increase in government spending. The fact that they do not is the college student's first clue that the economic model he is learning lacks realism. However, $\psi_{\mathbf{x}}(\mathbf{x},\mathbf{y})$, as output by **Economic Construction Set**, can be used to recommend government policy. Recall the definition of $\psi_{\mathbf{x}}(\mathbf{x},\mathbf{y})$:

$$\psi_{\mathbf{x}}(\mathbf{x},\mathbf{y}) = \begin{bmatrix} Y_g(\mathbf{x},\mathbf{y}) & Y_m(\mathbf{x},\mathbf{y}) & Y_r(\mathbf{x},\mathbf{y}) & Y_t(\mathbf{x},\mathbf{y}) \\ R_g(\mathbf{x},\mathbf{y}) & R_m(\mathbf{x},\mathbf{y}) & R_r(\mathbf{x},\mathbf{y}) & R_t(\mathbf{x},\mathbf{y}) \\ P_g(\mathbf{x},\mathbf{y}) & P_m(\mathbf{x},\mathbf{y}) & P_r(\mathbf{x},\mathbf{y}) & P_t(\mathbf{x},\mathbf{y}) \end{bmatrix}$$

If one divides the first row by the third row, the first three quotients represent the effects on real national income of a small increase in government spending, the issue of new money, and the tax rate, respectively. One should recommend that each of these actions be taken or not taken depending on the sign of their respective derivative. The second row is of more interest to people speculating on government bonds, though they also have to predict the direction of changes in g, m, and r to make use of this information. Of course, the derivatives in $\psi_{\mathbf{x}}(\mathbf{x},\mathbf{y})$ only have meaning if the model is well correlated with past data and it is that on which users of **Economic Construction Set** should concentrate.

College students are never explicitly introduced to the three-dimensional model used by **Economic Construction Set** and not until the graduate level do they get past two-dimensional IS-LM analysis. Then they no longer use the word "multiplier." Thus, if one is asked for their opinion on government policy, any discussion of the multiplier will probably be confused with its common, undergraduate meaning of $\psi'(g)$ from the one-dimensional model. Nevertheless, the analogy is a good one and it may help to clarify this section for undergraduate students who have seen the word "multiplier" used in their textbooks.

Once undergraduate students advance beyond learning that changes in national income are a constant multiple of changes in government spending, they study IS-LM analysis. Since they are still not provided with any mechanism for prices to adjust, they only solve F and G for Y and R, though the price level is sometimes discussed independently using what is called the AS-AD graph. To facilitate analysis, F and G are usually made to be linear functions of Y and R. When the subject of price adjustment is brought up at this level, students are told that this causes the level sets of F and G, what is called the IS and LM curves, to "shift." Discussion of shifting curves almost invariably means that a problem is being done in fewer dimensions than it is defined. However, if one has difficulty visualizing a three-dimensional system of equations, one can think of the IS-LM graph as a two-dimensional slice of Y, R, P space which is perpendicular to the P axis. The position of the slice is determined by the P to which the three-dimensional Newton's method algorithm described above converges. Thus, the slice intersects the level sets (surfaces in three-dimensional space) of F and G along different curves, so these curves appear to shift as P changes over time. These two curves form a cross in the first quadrant of the slice, their intersection being at the Y and R such that $\phi(\mathbf{x},\mathbf{y}) = 0$ for the given value of P. The curve leading away from the origin and describing a positive relation between Y and R is the level set of G and is called the LM curve. The curve describing a negative relation between Y and R is the level set of F and is called the IS curve. Thus, the simultaneous use of IS-LM analysis and AS-AD graphs is an adaptation of a three-dimensional problem to the textbook format of two-dimensional graphs, which has the (only) virtue of providing textbooks with lots of illustrations. It is my opinion that researchers should eschew what is essentially an expository technique in favor of just learning the math.

IV

Analysis of Inherent Errors in the Published Data

There are two sources of difficulty in writing numerical software: mistakes and errors. An example of a mistake is the assertion of mainstream economics that the real interest rate is the nominal interest rate minus the inflation rate. Fortunately, mistakes such as this can be easily excised from the soft sciences by any mathematician who cares to read their textbooks. Errors are more difficult to deal with and are divided into two types: round-off and inherent. Round-off errors result from the fact that computers cannot represent numbers exactly unless they are rational and their denominators are factors of the number system's base, usually two. This problem may be solved either by carrying more significant digits than are output or by using a factorial based number system, that is, 1, 2, 6, 24, 120,... rather than 1, 2, 4, 8, 16,.... Furthermore, algorithms can often be altered to reduce round-off error. An example of this is the implementation of equilibration and partial pivoting in the solution of systems of linear equations. Floating point arithmetic on modern personal computers, coupled with standard linear algebra algorithms, effectively eliminate round-off error for the relatively easy application of economics. To test this, **Economic Construction Set** was used to invert a 5×5 Hilbert matrix. The Hilbert matrix was then multiplied by its inverse and the result compared with the identity matrix. There were no discrepancies.

Inherent errors result from inaccuracies in the input data and cannot be eliminated but only estimated. All of the published historical data used by **Economic Construction Set** has four significant digits, so one cannot expect the output to have more accuracy. However, it may happen that a function magnifies error so that small changes in its input values result in large changes in its output values. At each time interval, **Economic Construction Set** accepts four historical values, g, m, r, and t, and returns three values describing the state of the economy, Y, R, and P. Because of difficulties in measuring the historical values, they are necessarily approximations of the actual values. All of the historical data used by **Economic Construction Set** are published by the federal government and given with four significant digits. It is the opinion of many economists, including some in the mainstream, that the accuracy of the federal government's data is greatly exaggerated. A few minutes of reflection on how economic data is actually collected tends to support this opinion. In particular, considerable controversy has centered on the reliability of information about the money supply. Besides the fact that the Federal Reserve purposely conceals new issues of money so that the government may spend it before people react by raising prices, it is uncertain exactly what qualifies as a medium of exchange. Less controversial but no more reliable are measurements of the interest rate. As anyone who has attempted to obtain a loan knows, the interest rate varies considerably with the credit history of the borrower. Furthermore, buyers of expensive items always have a choice between a low interest rate and a rebate, so it is very difficult to separate the price level from the interest rate. For instance, it is impossible for statisticians to quantify either the price level or the prevalent interest rate when car dealers advertise that one can either accept $1000 cash back or zero percent financing. Government spending is also in doubt because the bureaucracies which handle government expenditures are often motivated to conceal their operation in red tape. Furthermore, the measurement of time is inaccurate because government bureaucracies do not keep complete records of exactly when expenditures are made or of when taxes are received.

Nevertheless, let us generously grant the published data an accuracy within one part per 16384 (4.2144 significant digits), and determine the bounds on the error in the outputed values of Y, R, and P should the errors in g, m, r, and t combine in the most unfavorable way. The user should understand that the establishment of error bounds does not imply that the outputed results are actually that far off. It may happen that the errors cancel each other out so that the outputed values of Y, R, and P happen to be very close to their actual values. In fact, no assertion is being made about the probability density function which describes the actual location of Y, R, and P except that it is zero a certain distance from the predicted Y, R, and P. It is an old but still very wrong belief to assume that there is a "principle of indifference" which asserts that probability density functions which one knows nothing about are necessarily rectangular. To discuss the distribution of error within the established bounds would require precise knowledge of the sampling methods used by government statisticians, particularly whether they tend to overestimate or underestimate the data.

Recall from the previous section that the Taylor approximation adjusts Y, R, and P for the changes in g, m, r, and t at each time interval and that this approximation is better the smaller those changes are. For the changes in g, m, r, and t from one quarter to the next (A quarter is the longest allowable time interval), the Taylor approximation still requires refinement by Newton's method before the next quarter's Y, R, and P can be output. However, for the almost infinitesimal changes in g, m, r, and t due to error, the Taylor approximation is more than sufficient. Thus, the vector of errors in Y, R, and P is $\psi_{\mathbf{x}}(\mathbf{x},\mathbf{y})$ times the vector of errors in g, m, r, and t. The errors in g, m, r, and t are only known within certain bounds, however, so one cannot determine the exact errors in Y, R, and P but can only determine their error bounds by setting the errors in g, m, r, and t as large as possible, that is, at 1/16384th of their historical values. But these errors may be either positive or negative and the errors in Y, R, and P are each maximized by a different combination of signs on the input data. To avoid having to determine which combination of signs is needed to maximize each of the output values, one can take the absolute value of all the input data and of all the coefficients in $\psi_{\mathbf{x}}(\mathbf{x},\mathbf{y})$ before calculating the error bounds. This assures that no combination of signs on the errors in the input data could result in larger errors in the output data.

This is not the end of the story, however. $\psi_{\mathbf{x}}(\mathbf{x},\mathbf{y})$ is itself a function of g, m, r, and t and its coefficients are affected by any errors in them. Recall from the previous section that $\psi_{\mathbf{x}}(\mathbf{x},\mathbf{y})$ is the result of solving a linear system of equations, $-\phi_{\mathbf{y}}(\mathbf{x},\mathbf{y})\psi\mathbf{x}(\mathbf{x},\mathbf{y}) = \phi_{\mathbf{x}}(\mathbf{x},\mathbf{y})$. The derivatives of ϕ with respect to both $\mathbf{x}$ and $\mathbf{y}$ are known, but not perfectly because of inaccuracies in $\mathbf{x}$. Let us assume that the evaluation of the derivatives themselves do not magnify errors in their input values (this could only happen if the evaluations involve the subtraction of nearly equal quantities) so that all of the coefficients in both $\phi_{\mathbf{x}}(\mathbf{x,y})$ and $\phi_{\mathbf{y}}(\mathbf{x,y})$ are within 1/16384th of their calculated values. To condense the notation, we will consider the evaluation of each column of $\psi_{\mathbf{x}}(\mathbf{x,y})$ separately, as such:

$$\begin{aligned} A_{11}X_1 + A_{12}X_2 + A_{13}X_3 &= C_1 \\ A_{21}X_1 + A_{22}X_2 + A_{23}X_3 &= C_2 \\ A_{31}X_1 + A_{32}X_2 + A_{33}X_3 &= C_3 \end{aligned}$$

The uppercase C's represent a column of $\phi_{\mathbf{x}}(\mathbf{x,y})$ and the uppercase X's represent the associated column of $\psi\mathbf{x}(\mathbf{x,y})$. The uppercase A's represent coefficients of $\phi_{\mathbf{y}}(\mathbf{x,y})$. Let lowercase a's and c's represent the inherent error in those values so that, if better measuring devices were available, one would actually use $A_{11} + a_{11}$ instead of A_{11} and $C_1 + c_1$ instead of C_1, and so on. By hypothesis, all the data is accurate to within one part in 16384, so the bound on each error term is plus or minus its associated value divided by 16384. Lowercase x's will represent the resulting error in the X's. Construct the system of equations that should be solved and subtract the equations that are actually solved. The result is

$$\begin{aligned} (A_{11}+a_{11})x_1+(A_{12}+a_{12})x_2+(A_{13}+a_{13})x_3 &= c_1-(a_{11}X_1+a_{12}X_2+a_{13}X_3) \\ (A_{21}+a_{21})x_1+(A_{22}+a_{22})x_2+(A_{23}+a_{23})x_3 &= c_2-(a_{21}X_1+a_{22}X_2+a_{23}X_3) \\ (A_{31}+a_{31})x_1+(A_{32}+a_{32})x_2+(A_{33}+a_{33})x_3 &= c_3-(a_{31}X_1+a_{32}X_2+a_{33}X_3) \end{aligned}$$

All the lowercase a's are within 1/16384th of their associated uppercase A's so, if they all come out low, the square matrix is 16383/16384 times $\phi_{\mathbf{y}}(\mathbf{x,y})$. Because inversion can be distributed through scalar multiplication, the inverse matrix needed to calculate the error bounds is 16384/16383th of $\phi_{\mathbf{y}}(\mathbf{a},\mathbf{y}_0)^{-1}$ and it need not be calculated again. Separating the scalar multiples on the right-hand side yields

$$\begin{bmatrix} x_1 \\ x_2 \\ x_3 \end{bmatrix} = \frac{1}{16383} \begin{bmatrix} A_{11} & A_{12} & A_{13} \\ A_{21} & A_{22} & A_{23} \\ A_{31} & A_{32} & A_{33} \end{bmatrix}^{-1} \begin{bmatrix} C_1+A_{11}X_1+A_{12}X_2+A_{13}X_3 \\ C_2+A_{21}X_1+A_{22}X_2+A_{23}X_3 \\ C_3+A_{31}X_1+A_{32}X_2+A_{33}X_3 \end{bmatrix}$$

with the inverse already calculated except that all of its terms, and those of the column vector also, are taken to be positive regardless of their actual sign.

This calculation is repeated four times with different X's and C's to get the error-bounds for each column of $\psi_{\mathbf{x}}(\mathbf{x},\mathbf{y})$. Thus, associated with $\psi_{\mathbf{x}}(\mathbf{x},\mathbf{y})$ is the matrix $\begin{bmatrix} x_{11} & x_{12} & x_{13} & x_{14} \\ x_{21} & x_{22} & x_{23} & x_{24} \\ x_{31} & x_{32} & x_{33} & x_{34} \end{bmatrix}$ of error-bounds for each element of $\psi_{\mathbf{x}}(\mathbf{x},\mathbf{y})$. All of these error-bounds should be added to the absolute values of $\psi_{\mathbf{x}}(\mathbf{x},\mathbf{y})$ before using it to evaluate the Taylor approximation of the error-bounds for Y, R, and P. However, because multiplication distributes through addition, the vector of error-bounds for g, m, r, and t is first premultiplied by the absolute value of $\psi_{\mathbf{x}}(\mathbf{x},\mathbf{y})$ and then the same vector is premultiplied by the matrix of error-bounds associated with $\psi_{\mathbf{x}}(\mathbf{x},\mathbf{y})$ and, finally, the two products are added together. This method is algebraically equivalent to correcting $\psi_{\mathbf{x}}(\mathbf{x},\mathbf{y})$ for error before doing the Taylor approximation but it helps to reduce round-off error. The elements of the resulting vector are the error-bounds for Y, R, and P. When **Economic Construction Set** outputs a value for Y, R, and P at each time interval, these error-bounds are also output. The user should understand that the solution to the non-linear system of equations which should have been solved had better data been available is within plus or minus the error-bounds of the outputed values of Y, R, and P. Furthermore, he should understand that this error is inherent in the published data of the federal government and is not due to any weaknesses in the algorithm, of which there are none.

V

Keynesian and Monetarist Policy Recommendations

The multiplier is defined as

$$\psi_{\mathbf{x}}(\mathbf{x},\mathbf{y}) = \begin{bmatrix} Y_g(\mathbf{x},\mathbf{y}) & Y_m(\mathbf{x},\mathbf{y}) & Y_r(\mathbf{x},\mathbf{y}) & Y_t(\mathbf{x},\mathbf{y}) \\ R_g(\mathbf{x},\mathbf{y}) & R_m(\mathbf{x},\mathbf{y}) & R_r(\mathbf{x},\mathbf{y}) & R_t(\mathbf{x},\mathbf{y}) \\ P_g(\mathbf{x},\mathbf{y}) & P_m(\mathbf{x},\mathbf{y}) & P_r(\mathbf{x},\mathbf{y}) & P_t(\mathbf{x},\mathbf{y}) \end{bmatrix}$$

Recall from Section III that, if one divides the first row by the third row, the first three quotients represent the effects on real national income of a small increase in government spending, the issue of new money, and the tax rate, respectively. One should recommend that each of these actions be taken or not taken depending on the sign of their respective derivative. This is what is known as "fine-tuning" the economy. One nudges government spending, the issue of new money, and the tax rate in their appropriate directions, waits a few weeks to see what happens, recalculates the multiplier, and then nudges them again. Of the three output variables, Y, R, and P, economists watch R most closely since it takes time for statistics on Y and P to be gathered and they are not very accurate. The Federal Funds Rate exemplifies short-term interest rates because the loans are made for only a few hours and are between bankers who have little slack time in the response of their market. Until the late 1970's, this was virtually all there was to government policymaking. At that time, stagflation caused many economists to become disenchanted with this approach and they felt that domestic policy required a steady hand that was not swayed by the statistics of the day. This opinion is sometimes referred to as Monetarism, which is basically cautious Keynesianism. Monetarism attempts to hold the inputs, g, m, and r, steady at a (hopefully) optimal rate. Of the three inputs, only m requires close watching since g and r are controlled by lethargic bureaucracies while the money supply is at the command of one man, the Federal Reserve Chairman. Thus, the practical distinction between Keynesians and Monetarists is whether they ask the Federal Reserve Chairman to control ("peg", as it is called) R or m, respectively. Empirically, Keynesians won this argument because the money supply is a very inaccurate statistic compared to the Federal Funds Rate. In the late 1980's, economists returned to pegging interest rates, though the idea of stabilizing the inputs did not entirely die.

One solution to this dilemma is to make R an input variable and m an output variable, that is, define **y** = [Y, m, P] and **x** = [g, R, r, t]. Since it is Y/P, not R, that economists are most interested in, this is actually no more difficult than redefining **x** and **y**. Except for relabeling, the equations that follow would be unaffected by this change. Bond market speculators are less interested in the Federal Funds Rate than in long-term rates, which this model does not predict anyway, so they should not be concerned about switching R and m. However, such a switch would fly in the face of traditional IS-LM analysis, which clearly places R on the output side of the equation. Since I have my own theory of economics (see my book, **Axiomatic Theory of Economics**), it is not my place to advise mainstream economists on how to fix their theory. Users interested in pursuing this course should contact us (Aguilar Software Company) to have their software customized to reflect this change. For the remainder of this section we will ignore the empirical problems with measuring m and take the Monetarist position that the Federal Reserve Chairman should hold the money supply steady. The question is: Steady at what level?

Let the measure of success of an administration be the national wealth added during its four years in power. Wealth is the integral of real income, so we want to maximize the function $\int_0^{t_0} \frac{Y}{P} dt$ with t_0 a fixed amount of time, probably four years.

g and r are moved into **y** so **x** = [m, t] and W is a function of m only. This requires two more coordinate functions in ϕ, which the user will enter, denoted J(**x**,**y**) = 0 and K(**x**,**y**) = 0. Mainstream economics does not put any constraints on government spending or the tax rate, so these constraints must be purely political in nature. For instance, the ratio of government spending to national income represents the percentage of government involvement in the economy, with zero being pure free enterprise and unity being pure socialism. Mainstream economics considers this a political, not an economic, question and makes no distinction between government and private spending. Notice that the income identity, Y = C + I + g + X, gives equal weight to private spending, C + I + X, and government spending, g, without concerning itself that the government spending might be wasted. Thus, in keeping with the traditional separation of economics and politics, one could define J(**x**,**y**) (or K(**x**,**y**)) = b - $\frac{g}{Y}$, with b a constant representing the current political climate. Another possibility is the enactment of a balanced budget amendment so J(**x**,**y**) (or K(**x**,**y**)) = g - rY. Of course, the purpose of a central bank is to run deficits, so as long as a central bank exists it may be more reasonable to define J(**x**,**y**) (or K(**x**,**y**)) = b - (r+b)Y with b the ratio of the deficit to national income that voters will tolerate before demanding a reduction in government waste. One could also assume that bureaucratic growth is inexorable and make government spending, g, a function of time. Thus J(**x**,**y**) (or K(**x**,**y**)) = g - be^{kt} with b the rate of government spending at time zero and k a positive constant. This could be refined by making k a "reaction function" rather than a constant. Since there is political pressure to increase government spending during recessions (and, automatically, recessions cause more unemployment and welfare payments), k could increase as a recession deepens. The choice of coordinate functions is up to the user, but he must have two of them and they must be independent of each other. If he inputs two versions of the same relation, **Economic Construction Set** will require that they be input again.

Taking the derivative of W with respect to m requires an application of Leibnitz' rule. This requires that $\psi_{\mathbf{x}}(\mathbf{x},\mathbf{y})$ be continuous, which it should be if the user has not input any unusual functions for C, I, X, D, or A. The derivative is $W_m(m,\mathbf{y}) = \int_0^{t_0} \frac{PY_m - YP_m}{P^2} dt.$

W(m) is at a maximum where $W_m(m,\mathbf{y}) = 0$. Since W is a function of only one variable, **Economic Construction Set** maximizes it with the one-dimensional Newton's method algorithm that everyone is familiar with from first-semester calculus. The initial estimate is input by the user. Since each iteration is very time consuming, it is important that the user have a good idea what the optimal (constant) money supply is before beginning. If $W_m(m,\mathbf{y})$ is not approaching zero after a few iterations or the zero is a minimum of W(m), **Economic Construction Set** will ask for a new initial guess. The recommended money supply depends mostly on the price adjustment mechanism, A. If the user is doing traditional IS-LM analysis in which A = 1, the optimal money supply will be infinite and, after a few iterations, **Economic Construction Set** will inform the user that m seems to be increasing without bound.

Romberg integration is used to integrate $\frac{PY_m - YP_m}{P^2}$ from 0 to t_0, evaluating it every quarter. At each quarter $\phi(\mathbf{x},\mathbf{y}) = \begin{bmatrix} F(\mathbf{x},\mathbf{y}) \\ G(\mathbf{x},\mathbf{y}) \\ H(\mathbf{x},\mathbf{y}) \\ J(\mathbf{x},\mathbf{y}) \\ K(\mathbf{x},\mathbf{y}) \end{bmatrix} = 0$

This is solved for $\mathbf{y}$ = [Y, R, P, g, r] the same way that the three-dimensional problem is solved, as explained in Section III. More symbolic derivatives have to be calculated here, which adds overhead time and may require additional computer memory if one has a complicated economic model. Also, the five-dimensional Newton's method algorithm takes considerably longer to converge than the three-dimensional algorithm does. However, even for users who input very complicated coordinate functions, there should be no difficulty in solving five equations for five unknowns on a modern personal computer.

At the operational level, this software is addressed to high school honors students. Calculus of one variable is taught as a high school honors course and, at that level, multivariable calculus may seem very deep. However, if those students go on to study math or engineering, in two years they will realize that, in terms of theoretical difficulty, all of mainstream economics does not even compare to routine engineering problems. As one can see from reading this manual, mainstream economics is a fairly straightforward application of multivariable calculus. University students majoring in economics are only given an introduction to calculus of one variable, however. It does not include trigonometry and does not extend high-school calculus. This makes economics textbooks seem very convoluted as they attempt to talk around the three-dimensional problem. No effort is made at universities to solve the five-dimensional problem of optimizing the money supply. If the user already has some experience in college economics courses, he may be surprised that mainstream macroeconomics can be explained and considerably extended in a fairly short computer software manual.

From a purely number-crunching perspective, however, macroeconomic problems can strain large computers. Time and memory requirements increase factorially with the number of equations being solved, if it is done right, while the capabilities of computers only increase linearly with the money spent on them. Thus, when one hears about a computer model that solves hundreds of equations for hundreds of variables, one can assume that, even with the supercomputers available to the federal government, the equations are probably just thrown together without a great deal of theoretical rigor, or else they would have exceeded their computer's capabilities. Almost certainly the units of their equations do not work out (ratios are added to each other or to flows, for instance). Economists at that level attach a lot of importance to the adjective "computerized" and tend to construct models that are heavy on number crunching and light on theory. As complicated as they are, such models are always linear and there are limitations to what can be accomplished with a linear system of equations. That one is solving several hundred linear equations in as many unknowns does not address these limitations. **Economic Construction Set**, which has the ability to symbolically differentiate arbitrary, user-inputed functions, allows economists to address abrupt shifts in the economy. Thus, users of **Economic Construction Set** should not be intimidated by the fact that their analysis is "only done on a desktop computer." They can at least be assured that it is done right, even if it does not take into account popular statistics like consumer confidence, housing starts or applications for unemployment insurance. Besides overflowing the computer memory, throwing such statistics into **Economic Construction Set** would weaken the theory, not strengthen it.

Aguilar Software Company
7620 E. McKellips #424
Scottsdale, AZ 85257

www.axiomaticeconomics.com

BIBLIOGRAPHY

1) Robert Pirsig, **Zen and the Art of Motorcycle Maintenance**
New York, NY, Bantam Books, 1974, pg. 215,225

2) Albert Einstein and Leopold Infield, **The Evolution of Physics**
New York, NY, Simon and Schuster, 1938, pg. 27

3) Frederick Nietzsche, **Thus Spoke Zarathustra**
New York, NY, Penguin Books, 1954, pg. 261

4) Immanual Kant, **Critique of Pure Reason**
Garden City, NY, Doubleday & Company, Inc., 1966, pg xxxi

5) Willard Quine, **Methods of Logic**
Cambridge, MA, Harvard University Press, 1982, pg. 196

6) Halliday and Resnick, **Fundamentals of Physics**
New York, NY, John Wiley and Sons, 1988, pg. 274

7) Israel Kirzner, **The Economic Point of View**
Kansas City, KS, Sheed and Ward, Inc., 1976, pg 150

8) Hermann Hesse, **Siddhartha**
New York, NY, New Directions Publishing, 1951, pg. 26

9) Carl Menger, **Principles of Economics**
New York, NY, New York University Press, 1981, pg. 77

10) Robert Heilbroner and William Milberg,
The Crisis of Vision in Modern Economic Thought
Cambridge, MA, Cambridge University Press, 1995, pg. 9-10

11) Paul Ormerod, **The Death of Economics**
New York, NY, John Wiley and Sons, 1994, pg 93

12) David Hume, **On Human Nature and the Understanding**
Collier Classics in the History of Thought
New York, NY, Macmillan Publishing Co., 1962, pg. 50

13) Peter Griffin, **The Theory of Blackjack**
Davis, CA, Faculty Publishing, 1986, pg. 133

14) Paul Krugman, **The Accidental Theorist**
New York, NY, W. W. Norton & Co. Ltd., 1998, pg. 64

15) Robert Hall and John Taylor, **Macroeconomics**
New York, NY, W. W. Norton & Co. Ltd., 1991,
MacroSolve software (by Stephen King and Rick McConnell)

16) John Maynard Keynes,
The General Theory of Employment, Interest, and Money
New York, NY, Harcourt Brace Jovanovich, 1953, pg. 129

17) Archer Jones, **The Art of War in the Western World**
Barnes and Noble, Inc, 1987, pg. 45-46

18) Charles Kindleberger, **Manias, Panics and Crashes**
New York, NY, John Wiley & Sons, Inc., 1996, pg. 45

19) Charles Kindleberger, pg. 62

20) Heilbroner and Milberg, pg. 100

21) Paul Krugman, **The Self-Organizing Economy**
Malden, MA, Blackwell Publishers Inc., 1996, pg. 7

Index

T

U

W